Which process?

WHICH PROCESS?

An introduction to welding and related
processes and a guide to their selection

PETER HOULDCROFT

ABINGTON PUBLISHING

Woodhead Publishing Ltd in association with The Welding Institute
Cambridge England

Published by Abington Publishing
Woodhead Publishing Ltd, Abington Hall, Abington,
Cambridge CB1 6AH, England

First published 1990

© Woodhead Publishing Ltd

British Library Cataloguing in Publication Data
Houldcroft P T (Peter Thomas)
Which process?
1. Welding
I. Title
671.52

ISBN 1 85573 008 1

Designed by Geoff Green (text) and Chris Feely (cover), typeset by Goodfellow &
Egan, Cambridge and printed by St Edmundsbury Press, Bury St Edmunds

Preface

Many books on welding and joining give descriptions of joining processes and also indicate the types of application for which the processes are suitable. The information is, however, classified under headings such as process type e.g. arc, resistance or power beam. While this is satisfactory for a student or practitioner of joining it is not very helpful to an engineer or designer for whom the joining process is merely a means to an end. It is on the drawing board that the need for a connection is first perceived and the author has long wished to find a way of presenting information on joining processes which would allow its use at this point. This means reversing the normal method of presentation, a formidable task because of the numerous factors and qualifications which must be considered.

Of these factors the design of the joint and the type and thickness of the materials from which it is made seem instinctively to be of greatest significance. Starting from this point, and encouraged by John L Sanders, Director, Professional Affairs of The Welding Institute, a system has been devised in which processes are classified by usage in a manner not unlike the way Roget's Thesaurus classifies words according to usage. While the scheme does not attempt to go into the detail of joining procedures it does provide an initial judgement on feasibility and could suggest alternatives which might have been overlooked. It is hoped that the reader will also find the general advice on joining and the assessments of individual processes helpful. The aim throughout has been to reduce the amount of searching necessary to obtain information.

The author is indebted to John Loader, Head of Marketing and Information, The Welding Institute, for providing sources of information.

Peter Houldcroft
February 1990

THE ESAB GROUP

UK WELDER TRAINING CENTRE

The Esab Group Training Centre in Birmingham is recognised by many as the most advanced craft welding training centre in the United Kingdom.

It brings together highly qualified craft instructors with the latest designs of welding equipment and consumables produced by the largest manufacturer of welding products in the world.

With this strength behind it the Esab Group Training Centre offers welding craft training to the highest national and international standards in oxy acetylene, manual metal arc, tungsten inert gas and metal inert gas/metal active gas fusion welding processes.

Contact: **Les Ness**
Group Training Manager
Esab Group UK Ltd
Group Training Centre
Plume Street
Aston
Birmingham B6 7RU
Telephone: (021) 328 2711
Telefax: (021) 327 4341

Contents

Introduction

Objectives

This book has two objectives, firstly, to present a scheme to help those unfamiliar with the range of non-mechanical joining processes to make an initial judgement on which process is suitable for a particular joint or application and secondly, to provide an introduction to welding and related joining processes. Although it is intended primarily as a working system for the engineer it is hoped that the book will also prompt those who are more experienced to consider alternative solutions to their joining problems. No previous knowledge of joining technology is assumed as the application of the selection scheme starts on the drawing board, where the need for a connection is usually first perceived. The selection scheme indicates the feasibility of using a process and does not provide at this stage of development a detailed joining or welding procedure. Once the process is known, however, there are other sources of information, listed at the end of this section, which can be consulted for detailed procedures.

About the selection scheme

The two most important factors in selecting a welding or joining process are usually the design of the joint and the thickness of the material. The basis of the selection scheme is therefore a descriptive list of joints which together with their accompanying sketches cover almost all known types of connection. The choice of a joint type gives the first two digits of a three digit number. The third digit provides additional information, usually the thickness of the material to be joined. Not all the numbers available in the scheme have been used and the gaps have been left deliberately to provide scope for future development or to allow the scheme to be adapted to a particular user's requirements.

Having selected the three digit number from the first list (list A) on the basis of joint type and thickness this number is then looked up on list B. List B has alongside each number 28 welding and related processes arranged so that those on the left-hand side are suitable for welding large structural fabrications and site welding where the process must be taken to the job and those on the right-hand side are processes suitable for engineering components where the job is taken to the welding machine.

The dividing line is friction welding which is generally a process for engineering components but special forms exist which allow stud welding on structural fabrications. Accordingly processes such as Thermit, manual metal arc and submerged-arc appear to the left of friction welding and resistance and power beam welding appear to the right. An attempt is also made in list B to place those processes which have the least thermal effect on the workpiece to the right-hand side of each half of the page. Thus on the structural fabrication side the processes arc spot, plasma and percussion welding are further right than oxy-acetylene and TIG. On the engineering component side explosive welding, soldering, cold welding and adhesive bonding are to the right of flash and resistance welding.

List B may suggest to the enquirer that there are, for example, three candidate processes. The choice is finally narrowed down by consulting list C, an alphabetical list of joining processes in which they are pictured and their main characteristics are described together with their limitations and an indication of the factors affecting cost. It is in list C that the influence of material and service conditions is mentioned and the use of special techniques such as pulsed welding is covered.

List C also includes a literature reference which will enable the user of the selection scheme to examine any process in greater detail if necessary.

Joints and welds

It is welds which convert an assembled joint into a structural entity. This product of the joining process is still, however, called 'a joint'.

Joints in flat plate are fairly easily classified into types according to their shape. The usually accepted types are :- butt, lap, corner, T and edge (see Figure).

These terms are not sufficiently detailed to allow an adequate description of the more complex shapes of joint and list A includes 44 types of joint which are recognisably different.

There are several different types of weld used in these joints of which the two most common are butt and fillet. A butt weld which joins parts through their full thickness – a full penetration butt weld – is the strongest weld in tensile and fatigue as it allows lines of stress to pass through the joint smoothly. In joints made with partial penetration butt welds or with fillet welds the lines of stress are bent in passing through the joint, producing stress concentrations which reduce fatigue strength.

Edge preparation

When an arc or flame is used to make a weld the molten pool has a finite depth which depends on the heat input and speed of welding. As the thickness of the metal being joined is increased, this depth of penetration is unable to fuse the joint completely. Sometimes it will be possible to turn the joint over and complete the weld from the opposite side. Frequently, however, it is necessary to groove the joint to allow fusion to take place at the root. The weld may then be completed in a single or more usually a succession of passes of the arc or flame.

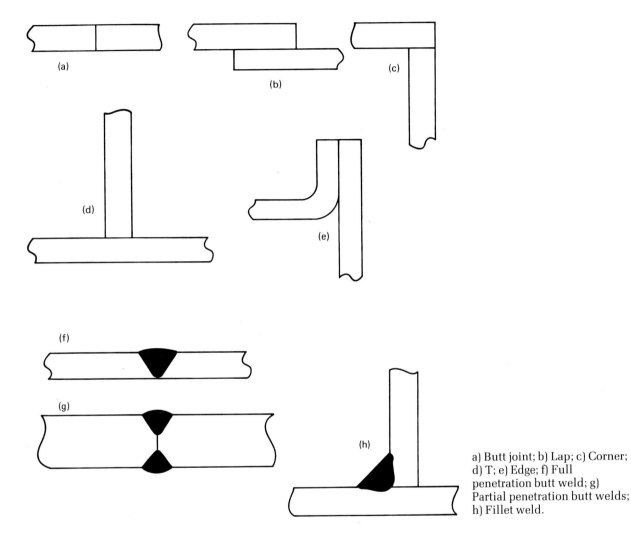

a) Butt joint; b) Lap; c) Corner; d) T; e) Edge; f) Full penetration butt weld; g) Partial penetration butt welds; h) Fillet weld.

Grooves in joints are of many different types, single chamfers or rounded grooves and may be single or double sided. The angle subtended by a groove must be sufficiently wide to allow weld metal to fuse the face of the joint without leaving a sharp re-entrant angle that might cause lack of fusion or result in trapped slag. It must also allow access for the welding electrode and vision of the root of the joint by the welder. Grooves are only rarely taken to a 'feather edge'. Edge preparations normally retain some of the original edge as a 'nose' or 'land' which not only makes it easier to control the root penetration but also makes it easier to assemble a joint ready for welding.

Backed welds

Although it is possible with skill to fuse the root of a joint completely by relying on surface tension to hold the molten metal in place, easier, faster welding is possible if the penetrating weld bead is supported in some way. There are five such ways:

1   Permanent or fusible backing in which a strip of the parent metal is placed under the joint and becomes fused into the weld;
2   Temporary backing using shaped bars, often of copper;
3   Flux backing using a trough filled with flux;
4   Root backing used with two sided welds in which only partial penetration is made on welding the first side; and
5   Gas backing using an inert gas shield to the penetration bead. This last type of backing is different from the previous four in that it is used to improve the shape of the penetration bead rather than to provide support which allows welding speed to be increased.

Properties required

Every joining method has its limitations because the material of the join itself is never exactly the same as the material in which it is placed. The material of the joint may be of different composition from the parent material (as with soldering) or it may be similar material in a different condition because of the effects of heat (as with a weld in an aluminium alloy). As a result, the joint as a whole may be subject to any one of a number of property limitations compared with the parent metal e.g. it might have reduced corrosion resistance, lower tensile strength, impaired fatigue or fracture toughness performance, lower resistance to the effects of temperature or may just be the wrong colour to match its surroundings. The selection of the joining process is therefore a compromise in which due note is taken of the intended use and service conditions of the joint. It is rare, however, for these considerations alone to decide the method of joining. It is also necessary to take account of costs and the availability of both welding plant and skills. Where high quality is a particular requirement it may be necessary to choose joint designs which allow access for non-destructive testing.

For a great many years only riveting or other mechanical joining methods were available and it is as well to recall that such methods were only gradually replaced by welding although they rarely developed the full strength of the material and were susceptible to crevice corrosion or other limitations. Mechanical joints, as well as joints made by soldering, brazing, adhesive bonding and resistance welding, require an overlap which adds to the total weight of the assembly compared with a simple butt joint. Surveys carried out at a time when welding was replacing riveting showed that welded structures were frequently 10 – 15% lighter than the riveted structures they replaced. Weight saving of this order is desirable in many applications and not merely those where weight is obviously important. It can even be worthwhile to save 10-15% of the construction weight of an expensive material.

The effects of heat

With joining methods such as welding, in which the material is melted and cast and that alongside the weld is subjected to heating and cooling, metallurgical changes can take place leading to a variety of weld defects. The literature on welding places a deal of emphasis on weld defects so it is as well to understand that the cause of most weld defects is known and they are avoidable with thought on the part of the designer and care by the operator or supervisor. Total freedom from all faults in welds is almost unattainable but many of the welds which are in service and performing adequately are known to be imperfect. The cost of a welding operation rises rapidly as demands for quality are increased so it is important to establish 'fitness-for-purpose' criteria before welding begins. Some welding defects are potentially more dangerous than others. Porosity which can occur in welds, soldered, brazed and adhesively bonded joints is the least potentially serious of defects. Cracks and also lack of fusion defects, however, are taken more seriously because they often lie in a plane across a direction of stress and being sharp they are potential initiators of fracture or fatigue.

Low carbon mild steel is the easiest material to weld by almost every process but as the carbon content or alloy content is increased to give higher strength the difficulties in welding increase. In like manner the problems in welding non-ferrous metals also tend to increase as tensile strength is raised by increasing the alloy content e.g. pure aluminium and its low alloys are easy to weld by a variety of methods but the high strength alloys typically used in aircraft manufacture are almost impossible to weld to the exacting standards of safety required for this application.

There are defects which arise in the weld metal and those which occur in the heat affected zone (HAZ). The defect most likely to occur in weld metal is 'hot' or solidification cracking usually found along the centreline of fusion welds. It can be caused in steels by a higher than normal sulphur content and in other alloys by not using the correct filler metal. With mechanised welding it can be the result of having a weld penetration which is too deep or by welding too fast, both of which allow the solidifying grains to grow together in such a way as to trap low melting point constituents.

One of the most common HAZ defects is HAZ hydrogen cracking which can occur in steels. It arises when hydrogen is present in a structure of a critical hardness. Sensitivity to this form of cracking increases with carbon and alloy content (carbon equivalent) and with weld cooling rate which is influenced by plate thickness and joint design. The preventative measures are to weld with low hydrogen (basic) electrodes and to reduce the cooling rate by preheating. Further help in deciding an appropriate procedure can be obtained from the British Standard 5135:1984 'Metal-arc welding of carbon-manganese steels'.

With welds in thick steel plate it is often necessary to ensure that the joints have adequate toughness as an assurance against brittle fracture. Brittle fracture can take place when three conditions are present, a suitable defect from which the crack can initiate, stress at this point of a suitable level, and a lack of toughness in the surrounding metal. It is a general rule with the manual metal arc (MMA) process to use basic electrodes for quality welds and for welds in thick plate.

Another defect which can occur in thick steel plate is lamellar tearing. This is a result of shortcomings in the plate itself accentuated by welding strains and an unsuitable joint design. Plate likely to be affected has poor properties, particularly toughness, in the through-thickness direction (sometimes called the Z direction). This is caused by zones of non-metallic inclusions lying in layers parallel to the surface which are opened up during welding to form step-like cracks. Modern steels are less affected by this defect than those made some years ago but if it is suspected that the plate might be susceptible to lamellar tearing joints should be designed so that welding stresses in the through-thickness direction are limited, e.g. by avoiding cruciform joints or heavy fillet welds. Another remedial action is to 'butter' the area likely to be affected before the connecting weld is made.

The application of heat in welding also results in distortion for which the treatment is inexact although there are a number of general rules. Much heat, particularly from a diffuse heat source such as the gas torch, causes the greatest distortion. The aim should always be to use the minimum of weld metal and to place it as symmetrically as possible. Some processes such as electron beam and laser (the power beam processes) are able to do this naturally and can allow welding of finished machined parts. Adhesive bonding, soldering, and to a lesser extent brazing, also allow distortion free joints because little or no heat is involved. Other solutions to the distortion problem are to preset parts at an angle before welding or to hold them rigidly in jigs. There are also methods for straightening parts after welding by thermal or mechanical means.

Costs

An indication is given in list C of the factors affecting costs, including an estimate of the cost of plant which, to take account of the effects of inflation, is given as a comparison with one of the most widely used processes, manual metal arc. Because the cost of welding plant varies widely (some plant such as that for electron beam being 50 or more times as costly as that for manual metal arc) it may be wondered why this can be justified. The answer is that such costly plant is used because it produces welds more quickly, with greater consistency or can make joints which cannot be made any other way. With many processes, particularly those which are basically mechanised, a substantial element in plant costs is usually the mechanical handling equipment rather than the welding plant itself. For certain applications such as pipe-welding there is specially developed automatic welding equipment. It is not always necessary for a manufacturer to purchase expensive welding plant for an unfamiliar process as there are now numerous specialist jobbing shops which can be used for trials or part production.

Additional information

The reader who wishes to follow up any of the topics mentioned above has an extensive literature to consult but introductions are given in:-

Hicks J G: *'A guide to designing welds'*. Publ Abington Publishing, 1989, 64 pages.
Gourd L M: *'Principles of welding technology'*. Publ Edward Arnold, 1980, 218 pages.
Lancaster L F: *'Metallurgy of welding'* Publ Allen and Unwin, 1987, 361 pages.
Houldcroft P and John R: *'Welding and cutting'*. Publ Woodhead-Faulkner, 1988, 232 pages.

Information on welding procedures is available as microcomputer software from The Welding Institute, Abington Hall, Abington, Cambridge CB1 6AL and INFOWELD, Wijtek Ltd, 128 Weirdale Ave, Whetstone, London N20 0AH.

How to use 'Which Process?'

First, find the joint number by selecting the joint which has to be made from the illustrations in LIST A. Many joints which look different are actually similar so this choice must be made carefully. This choice gives the first two digits in the joint number. The third digit indicates thickness and is a number from 0 to 4.

> **(Take as an example a butt weld in steel less than 1 mm thickness. This will have a number – 010)**

Second, now look up the joint number **(010)** on LIST B which gives the processes applicable to this joint and metal thickness.
> **(In this case OA, TIG, PA, PB, RSE.)**

Third, look up the processes applicable on LIST C. This list is a description of welding processes which also gives qualifications and notes on special techniques allowing a final choice to be made.

> **(For the example quoted LIST C indicates that OA, oxy-acetylene is a manual process giving considerable distortion; TIG, tungsten-arc can be used manually but is preferably mechanised; PA, plasma is as for TIG but gives better control on thinner material; PB, power beam is either electron beam or laser but that, of the two, laser welding is likely to be the preferred method; RSE, resistance seam welding normally requires an overlapped joint but in the thick-nesses in the example a special technique called 'mash-seam' allows a true butt weld to be made.)**

List A

Basic joint types

---

01   Butt, square edge, Sheet, plate and longitudinal in tube, single or double-sided.

0	thickness, mm	< 1
1		1–4
2		4–16
3		16–64
4		> 64

02   Butt, grooved. Sheet, plate and longitudinal in tube or pipe, single or double-sided.

0	thickness, mm	< 1
1		1–4
2		4–16
3		16–64
4		> 64

03   Butt in wire, rod or bar.

0	diameter, mm	< 1*
1		1–4
2		4–16
3		16–64
4		> 64

* Small sizes can be joined by a nugget of weld, braze metal or adhesive but this will be larger than diameter of work

04   Butt, circumferential. Tube up to 100 mm O/D.

0	diameter, mm	< 2*
1		2–8
2		> 8
3	non-rotatable, mm	2–8#
4		> 8

* When work can be rotated mechanised processes are best
# Also hollow sections

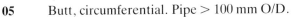

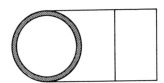

05 Butt, circumferential. Pipe > 100 mm O/D.

0	rotatable, mm	< 2 *
1		2–8
2		> 8
3	non-rotatable, mm	< 8
4		> 8

* When work can be rotated mechanised processes are best

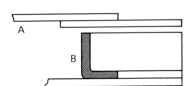

06 Butt in section.

0	thickness, mm	< 1
1		1–4
2		4–16
3		16–64
4		> 64

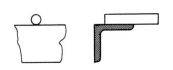

07 Lap in sheet and plate.*

0	thickness, mm	< 1
1		1–4
2		4–16
3		16–64
4		> 64

* Plug and slot welds can be made by MMA, TIG and MIG

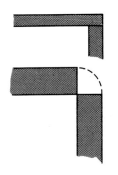

08 Lap, wire or rod to sheet or plate.

0	thickness, mm	< 1 equal*
1		1–4 equal
2		> 4 equal
3		1–4 plate #
4		cross wire

* Equal means wire diameter and other member similar thickness.
# Plate means wire to thicker member. These joints are poor in fatigue

09 Corner in sheet or plate.

0	thickness, mm	< 1*
1		1–4*
2		4–16
3		16–64
4		> 64

* Overlap joints can be made without filler by oxyacetylene, TIG and plasma in these thicknesses and in any thickness by power beam. Incomplete overlap used when process deposits filler metal, e.g. MMA and MIG. Folded and interlocked corners in sheet completed by brazing, soldering or adhesive bonding

10 Corner, flanged.

0	thickness, mm	< 1*
1		1–4
2		4–16
3		16–64
4		> 64

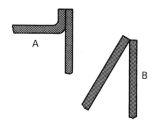

* Poor fit-up gives ragged fusion welds. Joint B unsuitable for welding by resistance seam, resistance spot or power beam

11 Corner, frame in bar, mitre or square.

0	thickness, mm	< 1*
1		1–4*
2		4–16#
3		16–64#
4		> 64#

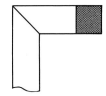

* Weld may be lumpy and require trimming
# Fusion welding will require edge preparation

12 Corner, frame in tube or hollow section.

0	thickness, mm	< 1
1		1–4
2		4–16
3		16–64
4		> 64

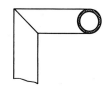

13 Corner, frame in section, mitre or square.

0	thickness, mm	< 1*
1		1–4
2		4–16
3		16–64
4		> 64

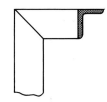

* Joints made by brazing, soldering or adhesive bonding in thin section will be lumpy and trimming will weaken. Full penetration welds can be safely trimmed

14 T in sheet or plate, fillet.*

0	thickness, mm	< 1
1		1–4
2		4–16
3		16–64
4		> 64

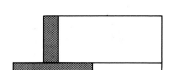

* T joints can be made by flanging the stem of the T when the joint resembles 07B

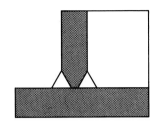

15   T in sheet or plate, full penetration.*

0	thickness, mm	< 1#
1		1–4
2		4–16
3		16–64
4		> 64

* Stem of T grooved for fusion welding.
  If T formed by two sections back-to-back joint resembles 01 or 02 with integral backing
Thicknesses of members approximately equal. See joint 44 for dissimilar thicknesses

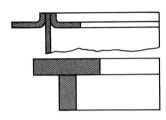

16   T in sheet, flanged.

0	thickness, mm	< 1
1		1–4

non-continuous, sheet and plate*

2	thickness, mm	1–4
3		16–64
4		> 64

* Welds can also be made from plate side using plug and slot welds with MMA, MIG, etc, and using spigot welds with B, TIG, S and AB. Arc welds require bevel on spigot. See joint 21

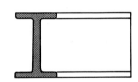

17   T in section, structural.

0	thickness, mm	< 1
1		1–4
2		4–16
3		16–64
4		> 64

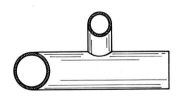

18   T in pipe, structural.

0	thickness, mm	< 1
1		1–4
2		4–16
3		16–64
4		> 64

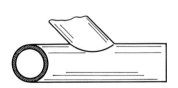

19   T in pipe, for flow.

0	thickness, mm	< 1
1		1–4
2		4–16
3		16–64
4		> 64

With large diameter drums mechanised welding can be easier from inside. See joint 26

20   Stud, boss and nozzle, set on.*

0	plate thickness, mm	< 1
1		1–4
2		4–16#
3		16–64#
4		> 64#

* With heavy joints plate must be sound to avoid lamellar tears. Friction welding not usually possible if set on member is a section. Flash welding requires upstand on plate or base part
\# Edge preparation may be necessary

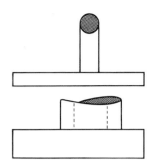

21   Stud, boss and nozzle, set in.*

0	plate thickness, mm	< 1
1		1–4
2		4–16#
3		16–64#
4		> 64#

* May be welded from either side
\# Bevelled spigot required for arc welds

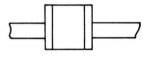

22   Stud, boss and nozzle, set through.

0	plate thickness, mm	< 1
1		1–4
2		4–16*
3		16–64*
4		> 64*#

* Edge preparation may be necessary
\# Plate must be sound–avoid lamellar tears

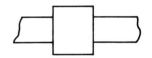

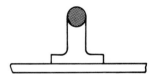

23   Stud, boss & nozzle, flanged, set on.*

0	plate thickness, mm	< 1
1		1–4
2		4–16
3		16–64
4		> 64

* With heavy joints plate must be sound to avoid lamellar tears

24   Stud, boss and nozzle, flanged, set in.*

0	plate thickness, mm	< 1
1		1–4
2		4–16#
3		16–64#
4		> 64#

* Usually superior in fatigue–joints 21, 23 or 25.
\# Edge preparation necessary except for power beam

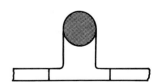

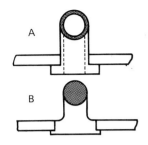

25   Stud, boss and nozzle, flanged, set through.

0   plate thickness, mm   < 1*
1   1–4
2   4–16#
3   16–64#
4   > 64#

* Resistance projection feasible with A and power beam with B
# Edge preparation may be necessary

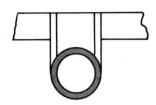

26   Tube/tubesheet or flange, front face.

0   tube thickness, mm   < 1
1   1–2
2   2–4*
3   4–8*
4   > 8*

* Edge preparation, usually by recessing tube, may be required

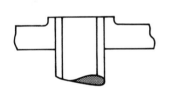

27   Tube/tubesheet or flange, front face trepanned.

0   tube thickness, mm   < 1
1   1–2
2   2–4*
3   4–8*
4   > 8*

* Not usually applicable

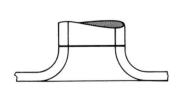

28   Tube/tubesheet or flange, back face.*

0   tube thickness, mm   < 1
1   1–2
2   2–4
3   4–8#
4   > 8#

* These joints are welded from inside the bore with special equipment
# Not usually applicable to tubeplates
   A only for friction and flash

29   Tube/tubesheet* or flange, formed or machined.

0   tube thickness, mm   < 1*
1   1–2*
2   2–4#
3   4–8#
4   > 8#

* Tubesheet pierced and formed
# Edge preparation may be necessary
   Check for access when welding tube/tubeplates

30 Tube closure, disc and corner weld.*

0	tube thickness, mm	< 1
1		1–2
2		2–4
3		4–8
4		> 8

* Compare joint 09. Mechanised welding best. Disc may be recessed for location

31 Tube closure, disc and flange.*

0	tube thickness, mm	< 1
1		1–2
2		2–4
3		4–8
4		> 8

* Compare joint 10. Mechanised welding best

32 Tube closure, cap, exterior.

0	tube thickness, mm	< 1
1		1–2
2		2–4
3		4–8
4		> 8

33 Tube closure, cap, interior, inwards.

0	tube thickness, mm	< 1
1		1–2
2		2–4
3		4–8
4		> 8

34 Tube closure, cap, interior, outwards.

0	tube thickness, mm	< 1
1		1–2
2		2–4#
3		4–8#
4		> 8#

Edge preparation may be necessary for arc welds

35 Tube closure, plug and fillet.

0	tube thickness, mm	< 1
1		1–2
2		2–4
3		4–8
4		> 8

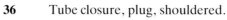

36   Tube closure, plug, shouldered.

0	tube thickness, mm	< 1
1		1–2
2		2–4
3		4–8
4		> 8

37   Tube closure, plug, shaped.

0	tube thickness, mm	< 1
1		1–2
2		2–4
3		4–8
4		> 8

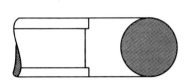

38   Tube closure, plug, butted.*

0	tube thickness, mm	< 1
1		1–2
2		2–4#
3		4–8#
4		> 8#

* Small spigot for purposes of location only
# Bevel plug for arc processes

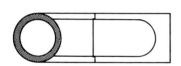

39   Tube closure, plug, butted, recessed.*

0	tube thickness, mm	< 1
1		1–2
2		2–4
3		4–8
4		> 8

* Small lip may be necessary for purposes of locating parts

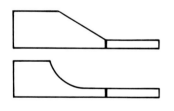

40   Dissimilar thickness, butt, tapered.

0	thickness, mm	< 1*
1		1–2
2		2–4
3		4–8
4		> 8

* Thickness refers to thinner member
  Smooth shape gives easier welding and better fatigue properties

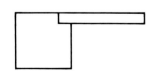

41   Dissimilar thickness, butt, rebated.

0	thickness, mm	< 1*
1		1–2
2		2–4#
3		4–8#
4		> 8#

* Thickness refers to thinner member
# Bevel thick member for arc welds
Requires preheat when arc welding except with pulsed TIG

42   Dissimilar thickness, fillet or lap.

0	thickness, mm	< 1*
1		1–2
2		2–4
3		4–8
4		> 8

* Thickness refers to thinner member
Requires preheat when arc welding except with pulsed TIG

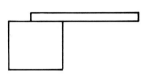

43   Dissimilar thickness, T. *

0	thickness, mm	< 1
1		1–2
2		2–4
3		4–8
4		> 8

* Thickness refers to thinner member
Requires preheat when arc welding except with pulsed TIG

44   Surfacing.

0	thickness, mm	< 1*
1		1–2
2		2–4
3		4–8
4		> 8

* Thickness refers to clad layer

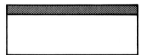

electric

011.
012
021
22
23
41
42
43
44

010
21
22
43

List B

Applicability of processes to joints shown in list A

● indicates that the process is applicable
· indicates process not applicable

Joint No.											Process																
	T	ES	OA	B	MMA	SA	EG	FC	MIG	AST	TIG	ASP	PA	PE	FR	DF	MB	F	RP	RSE	RSP	RU	PB	US	EX	S	CP	AB
010	·	·	●	·	·	·	·	·	·	·	●	·	●	·	·	·	·	·	·	●	·	·	●	·	·	·	●	·
011	·	·	●	·	●	·	·	·	●	·	●	·	●	·	·	·	·	·	·	●	·	●	●	·	·	·	·	·
012	·	·	·	·	●	●	●	●	●	·	·	·	·	·	·	·	·	●	·	·	·	●	●	·	·	·	·	·
013	●	●	·	·	·	●	●	●	●	·	·	·	·	·	·	·	·	●	·	·	·	·	●	·	·	·	·	·
014	●	●	·	·	·	·	·	·	·	·	·	·	·	·	·	·	·	·	·	·	·	·	●	·	·	·	·	·
020	See 010																											
021	·	·	●	·	●	●	·	·	●	·	●	·	·	·	·	·	·	·	·	·	·	·	·	·	·	·	·	·
022	·	·	●	·	●	●	·	·	●	·	●	·	·	·	·	·	·	·	·	·	·	·	·	·	·	·	·	·
023	·	·	·	·	●	●	·	·	●	·	●	·	·	·	·	·	·	·	·	·	·	·	·	·	·	·	·	·
024	·	·	·	·	●	●	●	●	●	·	·	·	·	·	·	·	·	·	·	·	·	·	·	·	·	·	·	·
030	·	·	●	·	·	·	·	·	·	·	·	·	●	●	·	·	·	·	·	·	·	●	●	·	·	·	●	·
031	·	·	●	·	·	·	·	·	·	·	·	·	●	●	●	·	·	·	·	·	·	●	●	·	·	·	●	·
032	·	·	·	·	·	·	·	·	·	·	·	·	·	●	●	·	●	·	·	·	·	●	●	·	·	·	●	·
033	●	·	·	·	·	·	·	·	·	·	·	·	·	●	●	·	●	·	·	·	·	●	●	·	·	·	●	·
034	·	·	·	·	·	·	·	·	·	·	·	·	·	●	●	·	●	·	·	·	·	●	●	·	·	·	●	·
040	·	·	·	·	·	·	·	·	·	·	●	·	·	●	·	●	●	·	·	·	·	●	·	·	·	·	●	·
041	·	·	·	·	·	·	·	●	●	·	●	·	●	·	●	·	·	●	·	·	·	·	●	·	·	·	●	·
042	·	·	·	·	·	·	·	●	●	·	●	·	·	●	·	·	·	●	·	·	·	·	●	·	·	·	●	·
043	·	·	●	·	●	·	·	·	●	·	●	·	·	·	·	●	●	●	·	·	·	·	●	·	·	·	●	·
044	·	·	·	·	●	·	·	●	●	·	●	·	·	·	·	·	●	●	·	·	·	·	●	·	·	·	●	·

T Thermit, ES electroslag, OA oxyacetylene, B brazing, MMA manual metal arc, SA submerged-arc, EG electrogas, FC flux-cored arc, MIG gas metal arc, AST arc stud, TIG gas tungsten arc, ASP arc spot, PA plasma, PE percussion, FR friction, DF diffusion bonding, MB MIAB, F flash, RP resistance projection, RSE resistance seam, RSP resistance spot, RU resistance upset, PB power beam, US ultrasonic, EX explosive, S soldering, CP cold pressure, AB adhesive bonding. Processes to the right of FR are not generally applicable to structural fabrications or site welding. Processes towards the right of each half of each line can have less thermal affect on the workpiece.

30 Which Process?

Joint No.	T	ES	OA	B	MMA	SA	EG	FC	MIG	AST	TIG	ASP	PA	PE	FR	DF	MB	F	RP	RSE	RSP	RU	PB	US	EX	S	CP	AB
050	·	·	·	·	·	·	·	·	·	·	●	·	●	·	·	·	·	·	·	·	·	·	●	·	·	·	·	·
051	·	·	·	·	·	●	●	●	●	●	●	·	●	·	●	·	·	●	·	·	·	·	●	·	·	·	·	·
052	·	·	·	·	·	●	●	●	●	·	·	·	·	·	●	·	·	●	·	·	·	·	●	·	·	·	·	·
053	·	·	●	·	●	·	·	●	●	·	●	·	·	·	·	●	·	●	·	·	·	·	●	·	·	·	·	·
054	·	·	·	·	●	·	·	●	●	·	●	·	·	·	·	·	·	●	·	·	·	·	●	·	·	·	·	·
060	·	·	●	·	·	·	·	·	·	·	●	·	·	·	·	·	·	·	·	·	·	·	·	·	·	·	·	·
061	·	·	●	·	●	·	·	·	●	·	●	·	●	·	·	·	·	●	·	·	·	·	·	·	·	·	·	·
062	·	·	·	·	●	·	·	●	●	·	·	·	·	·	●	·	·	●	·	·	·	·	·	·	·	·	·	·
063	●	·	·	·	●	·	·	●	●	·	·	·	·	·	●	·	·	●	·	·	·	·	·	·	·	·	·	·
064	●	·	·	·	●	·	·	●	●	·	·	·	·	·	●	·	·	●	·	·	·	·	·	·	·	·	·	·
070	·	·	·	●	●	·	·	·	·	·	●	●	●	·	·	·	·	·	●	●	●	●	●	●	·	●	●	●
071	·	·	●	●	●	·	·	·	·	·	●	●	●	·	·	●	·	·	●	●	●	·	●	·	·	●	·	●
072	·	·	·	●	●	●	·	●	●	·	●	●	·	·	·	●	·	·	·	·	●	·	●	·	·	·	·	·
073	·	·	·	●	●	●	·	●	●	·	·	·	·	·	·	●	·	·	·	·	·	·	●	·	·	·	·	·
074	·	·	·	●	●	●	●	●	●	·	·	·	·	·	·	·	·	·	·	·	·	·	●	·	·	·	·	·
080	·	·	·	●	·	·	·	·	·	·	●	·	●	·	·	·	·	·	·	·	·	·	·	·	●	●	·	●
081	·	·	●	●	●	·	·	·	·	·	●	·	·	·	·	·	·	·	●	·	·	·	·	·	·	●	·	●
082	·	·	●	●	●	·	·	●	●	·	·	·	·	·	·	·	·	·	·	·	·	·	●	·	·	·	·	·
083	·	·	·	●	●	·	·	●	●	·	●	·	·	·	·	·	·	·	·	·	·	·	●	·	·	·	·	●
084	·	·	·	●	●	·	·	·	·	·	●	·	●	·	·	·	·	·	●	·	·	·	·	·	·	·	·	·
090	·	·	·	●	·	·	·	·	·	·	●	·	●	·	·	·	·	·	·	·	·	·	●	·	·	●	·	·
091	·	·	●	●	●	●	·	·	●	·	●	·	●	·	·	·	·	·	·	·	·	·	●	·	·	●	·	·
092	·	·	·	●	●	●	·	·	●	·	·	·	·	·	·	·	·	·	·	·	·	·	●	·	·	·	·	·
093	·	·	·	●	●	●	●	·	●	·	·	·	·	·	·	·	·	·	·	·	·	·	●	·	·	·	·	·
094	·	●	·	●	●	●	·	·	·	·	·	·	·	·	·	·	·	·	·	·	·	·	●	·	·	·	·	·
100	·	·	●	●	·	·	·	·	·	·	●	·	●	·	·	·	·	·	·	●	●	·	●	●	·	●	●	●
101	·	·	●	●	·	·	·	·	·	·	●	●	●	·	·	·	·	·	·	·	·	·	●	·	·	·	·	●
102	·	·	·	●	●	●	·	●	●	·	·	·	·	·	·	·	·	·	·	·	·	·	●	·	·	·	·	·
103	Not an appropriate joint for this thickness																											
104	Not an appropriate joint for this thickness																											
110	·	·	●	●	·	·	·	·	·	·	●	·	●	●	·	·	·	·	·	·	·	·	●	·	·	●	·	·
111	·	·	●	●	·	·	·	·	·	·	●	·	●	●	·	·	·	·	·	·	·	·	●	·	·	·	·	·
112	·	·	·	●	·	·	·	·	·	·	●	●	·	·	●	·	●	·	●	·	·	·	·	·	·	·	·	·
113	·	·	·	●	●	·	·	·	●	·	·	·	·	·	●	·	·	●	·	·	·	·	·	·	·	·	·	·
114	●	·	·	●	·	·	·	·	●	·	·	·	·	·	●	·	·	●	·	·	·	·	·	·	·	·	·	·

T Thermit, ES electroslag, OA oxyacetylene, B brazing, MMA manual metal arc, SA submerged-arc, EG electrogas, FC flux-cored arc, MIG gas metal arc, AST arc stud, TIG gas tungsten arc, ASP arc spot, PA plasma, PE percussion, FR friction, DF diffusion bonding, MB MIAB, F flash, RP resistance projection, RSE resistance seam, RSP resistance spot, RU resistance upset, PB power beam, US ultrasonic, EX explosive, S soldering, CP cold pressure, AB adhesive bonding. Processes to the right of FR are not generally applicable to structural fabrications or site welding. Processes towards the right of each half of each line can have less thermal affect on the workpiece.

Joint No.	T	ES	OA	B	MMA	SA	EG	FC	MIG	AST	TIG	ASP	PA	PE	FR	DF	MB	F	RP	RSE	RSP	RU	PB	US	EX	S	CP	AB
120	·	·	·	●	·	·	·	·	·	·	·	·	●	·	·	·	●	●	·	·	·	·	·	·	·	●	·	·
121	·	·	●	●	●	·	·	·	●	·	●	·	·	·	·	·	●	●	·	·	·	·	·	·	·	·	·	·
122	·	·	·	●	·	·	·	·	●	·	●	·	·	·	·	·	●	·	·	·	·	·	·	·	·	·	·	·
123	·	·	·	·	·	·	●	●	·	·	·	·	·	·	·	·	●	·	·	·	·	·	·	·	·	·	·	·
124	·	·	·	●	·	·	●	●	●	·	·	·	·	·	·	·	●	·	·	·	·	·	·	·	·	·	·	·
130	·	·	●	●	·	·	·	·	·	·	●	·	·	·	·	·	·	·	·	·	·	·	·	·	·	●	·	·
131	·	·	●	●	●	·	·	·	●	·	●	·	·	·	·	·	●	·	·	·	·	·	·	·	·	·	·	·
132	·	·	·	●	·	·	·	·	●	·	·	·	·	·	·	·	●	·	·	·	·	·	·	·	·	·	·	·
133	·	·	·	●	·	·	·	·	●	·	·	·	·	·	·	·	●	·	·	·	·	·	·	·	·	·	·	·
134	●	·	·	●	·	·	·	·	●	·	·	·	·	·	·	·	●	·	·	·	·	·	·	·	·	·	·	·
140	·	·	·	●	·	·	·	·	●	·	·	·	·	·	·	·	·	·	·	·	·	·	·	·	·	●	·	●
141	·	·	●	●	·	·	·	·	●	·	·	·	·	·	·	·	·	·	·	·	·	·	·	·	·	·	·	·
142	·	·	·	·	●	●	●	●	●	·	·	·	·	·	·	·	·	·	·	·	·	·	·	·	·	·	·	·
143	·	·	·	·	●	●	●	●	●	·	·	·	·	·	·	·	·	·	·	·	·	·	·	·	·	·	·	·
144	·	·	·	·	●	●	●	●	●	·	·	·	·	·	·	·	·	·	·	·	·	·	·	·	·	·	·	·
150	·	·	·	●	·	·	·	·	●	·	·	·	·	·	·	·	·	·	·	·	·	·	·	·	·	●	·	●
151	·	·	·	●	·	·	·	·	●	·	·	·	·	·	·	·	·	●	·	·	●	·	·	·	·	·	·	·
152	·	·	·	·	●	●	●	●	●	·	·	·	●	·	·	·	·	●	·	·	●	·	·	·	·	·	·	·
153	·	·	·	·	●	●	●	●	●	·	·	·	●	·	·	·	·	●	·	·	●	·	·	·	·	·	·	·
154	●	●	·	·	●	●	●	●	●	·	·	·	●	·	·	·	·	●	·	·	●	·	·	·	·	·	·	·
160	·	·	●	●	·	·	·	·	·	·	●	·	●	·	·	·	·	·	·	·	·	●	·	·	·	●	·	●
161	·	·	●	●	●	·	·	·	●	·	●	·	·	·	·	·	·	·	·	·	·	·	·	·	·	·	·	·
162	·	·	·	●	●	·	·	·	●	·	●	●	·	·	·	·	·	·	·	●	·	·	·	●	·	●	·	●
163	·	·	·	·	·	·	·	·	·	·	·	·	·	·	·	·	·	·	·	●	·	·	·	·	·	·	·	·
164	Not an appropriate joint for this thickness																											
170	·	·	·	●	·	·	·	·	·	·	●	·	●	·	·	·	·	·	·	·	·	·	·	·	·	●	·	·
171	·	·	·	·	●	·	·	·	●	·	●	·	·	·	·	·	·	·	·	·	·	·	·	·	·	·	·	·
172	·	·	·	·	●	·	·	●	●	·	·	·	·	·	·	·	·	·	·	·	·	·	·	·	·	·	·	·
173	·	·	·	·	●	·	·	●	●	·	·	·	·	·	·	·	·	·	·	·	·	·	·	·	·	·	·	·
174	·	●	·	·	●	·	·	●	●	·	·	·	·	·	·	·	·	·	·	·	·	·	·	·	·	·	·	·
180	·	·	·	●	·	·	·	·	·	·	●	·	·	·	·	·	·	·	·	·	·	·	·	·	·	●	·	·
181	·	·	·	·	●	·	·	·	●	·	●	·	●	·	·	·	·	·	·	·	·	·	·	·	·	·	·	·
182	·	·	·	·	●	·	·	●	●	·	·	·	·	·	·	·	·	·	·	·	·	·	·	·	·	·	·	·
183	·	·	·	·	●	·	·	●	●	·	·	·	·	·	·	·	·	·	·	·	·	·	·	·	·	·	·	·
184	·	·	·	·	●	·	·	●	●	·	·	·	·	·	·	·	·	·	·	·	·	·	·	·	·	·	·	·

T Thermit, ES electroslag, OA oxyacetylene, B brazing, MMA manual metal arc, SA submerged-arc, EG electrogas, FC flux-cored arc, MIG gas metal arc, AST arc stud, TIG gas tungsten arc, ASP arc spot, PA plasma, PE percussion, FR friction, DF diffusion bonding, MB MIAB, F flash, RP resistance projection, RSE resistance seam, RSP resistance spot, RU resistance upset, PB power beam, US ultrasonic, EX explosive, S soldering, CP cold pressure, AB adhesive bonding. Processes to the right of FR are not generally applicable to structural fabrications or site welding. Processes towards the right of each half of each line can have less thermal affect on the workpiece.

32 Which Process?

Joint No.	T	ES	OA	B	MMA	SA	EG	FC	MIG	AST	TIG	ASP	PA	PE	FR	DF	MB	F	RP	RSE	RSP	RU	PB	US	EX	S	CP	AB
190				●									●													●		
191					●				●		●		●															
192				●					●		●																	
193					●			●	●																			
194					●			●	●																			
200				●										●												●		
201				●	●					●				●	●				●									
202			●					●	●						●	●		●	●									
203					●				●						●	●		●										
204					●				●						●	●		●										
210				●							●		●										●			●		●
211			●						●		●												●			●		●
212			●						●		●												●			●		●
213					●			●	●														●					
214					●			●	●														●					
220			●										●													●		●
221			●						●																	●		●
222			●						●																	●		●
223					●	●		●	●																			
224					●	●		●	●																			
230				●					●		●			●					●		●			●		●		●
231			●	●	●				●		●	●							●		●					●		●
232			●	●	●				●		●															●		●
233					●				●																			
234					●	●		●	●																			
240														●									●					
241			●						●		●			●									●					
242					●	●		●	●		●												●					
243					●	●		●	●														●					
244					●	●		●	●														●					
250			●											●					●				●			●		●
251			●	●							●								●				●			●		●
252				●	●			●			●												●					
253					●			●	●														●					
254					●			●	●														●					

T Thermit, ES electroslag, OA oxyacetylene, B brazing, MMA manual metal arc, SA submerged-arc, EG electrogas, FC flux-cored arc, MIG gas metal arc, AST arc stud, TIG gas tungsten arc, ASP arc spot, PA plasma, PE percussion, FR friction, DF diffusion bonding, MB MIAB, F flash, RP resistance projection, RSE resistance seam, RSP resistance spot, RU resistance upset, PB power beam, US ultrasonic, EX explosive, S soldering, CP cold pressure, AB adhesive bonding. Processes to the right of FR are not generally applicable to structural fabrications or site welding. Processes towards the right of each half of each line can have less thermal affect on the workpiece.

Joint No.	Process																											
	T	ES	OA	B	MMA	SA	EG	FC	MIG	AST	TIG	ASP	PA	PE	FR	DF	MB	F	RP	RSE	RSP	RU	PB	US	EX	S	CP	AB
260				●							●		●										●		●	●		●
261			●	●							●		●										●		●	●		●
262			●	●	●						●												●		●	●		●
263					●				●		●												●					
264					●			●	●														●					
270											●																	
271											●																	
272	Not an appropriate joint for this thickness																											
273	Not an appropriate joint for this thickness																											
274	Not an appropriate joint for this thickness																											
280												●																
281											●																	
282											●				●		●											
283					●			●			●				●		●											
284					●			●			●				●		●											
290												●					●						●					
291		●										●					●						●					
292		●									●	●			●		●						●					
293				●				●			●				●		●						●					
294				●				●			●				●		●						●					
300			●	●								●														●		
301			●	●							●	●														●		
302		●									●								●									
303					●	●			●		●												●					
304					●	●		●	●														●					
310			●	●							●		●						●	●	●		●	●		●	●	●
311		●		●							●		●						●	●	●		●	●		●	●	●
312			●	●	●				●		●												●					
313					●	●			●														●					
314					●	●		●	●														●					
320			●																							●		●
321				●							●	●														●		●
322			●	●	●				●		●																	
323					●			●	●																			
324					●	●		●	●																			

T Thermit, ES electroslag, OA oxyacetylene, B brazing, MMA manual metal arc, SA submerged-arc, EG electrogas, FC flux-cored arc, MIG gas metal arc, AST arc stud, TIG gas tungsten arc, ASP arc spot, PA plasma, PE percussion, FR friction, DF diffusion bonding, MB MIAB, F flash, RP resistance projection, RSE resistance seam, RSP resistance spot, RU resistance upset, PB power beam, US ultrasonic, EX explosive, S soldering, CP cold pressure, AB adhesive bonding. Processes to the right of FR are not generally applicable to structural fabrications or site welding. Processes towards the right of each half of each line can have less thermal affect on the workpiece.

34 Which Process?

Joint No.	T	ES	OA	B	MMA	SA	EG	FC	MIG	AST	TIG	ASP	PA	PE	FR	DF	MB	F	RP	RSE	RSP	RU	PB	US	EX	S	CP	AB
330				•							•		•										•			•		•
331			•	•							•		•										•					•
332			•	•	•	•			•		•												•			•		•
333				•	•	•		•	•		•												•					
334					•	•		•															•					
340				•									•										•			•		•
341			•	•							•		•							•	•		•			•		•
342			•	•	•				•		•									•	•		•			•		•
343				•	•	•		•	•		•												•					
344				•	•	•		•	•														•					
350				•							•												•			•		•
351				•							•		•										•			•		•
352				•	•				•		•												•			•		•
353				•	•	•		•	•														•					•
354				•	•	•		•	•														•					
360				•									•										•			•		•
361				•							•												•			•		•
362				•	•			•			•												•			•		•
363				•	•	•		•	•		•												•					
364					•			•	•														•					
370				•							•		•										•			•		•
371				•							•		•										•			•		•
372				•	•			•			•								•				•					
373					•			•											•				•					
374					•			•															•					
380	Use joint 360																											
381																							•					
382					•			•			•				•		•						•					
383					•			•			•				•		•						•					
384					•			•							•		•						•					
390														•		•							•					
391											•					•	•						•					
392											•				•		•	•					•					
393					•			•							•		•						•					
394				•				•							•		•						•					

T Thermit, ES electroslag, OA oxyacetylene, B brazing, MMA manual metal arc, SA submerged-arc, EG electrogas, FC flux-cored arc, MIG gas metal arc, AST arc stud, TIG gas tungsten arc, ASP arc spot, PA plasma, PE percussion, FR friction, DF diffusion bonding, MB MIAB, F flash, RP resistance projection, RSE resistance seam, RSP resistance spot, RU resistance upset, PB power beam, US ultrasonic, EX explosive, S soldering, CP cold pressure, AB adhesive bonding. Processes to the right of FR are not generally applicable to structural fabrications or site welding. Processes towards the right of each half of each line can have less thermal affect on the workpiece.

Joint No.	T	ES	OA	B	MMA	SA	EG	FC	MIG	AST	TIG	ASP	PA	PE	FR	DF	MB	F	RP	RSE	RSP	RU	PB	US	EX	S	CP	AB
400	·	·	·	·	·	·	·	·	·	·	●	·	·	·	·	·	·	·	·	·	·	·	●	·	·	·	·	·
401	·	·	●	·	●	·	·	·	●	·	●	·	·	·	·	·	·	·	·	·	·	·	●	·	·	·	·	·
402	·	·	·	·	●	·	·	·	●	·	●	·	·	·	·	·	●	·	·	·	·	·	●	·	·	·	·	·
403	·	·	·	·	●	●	·	●	●	·	·	·	·	·	·	·	●	·	·	·	·	·	·	·	·	·	·	·
404	·	·	·	·	●	●	·	●	●	·	·	·	·	·	·	·	●	·	·	·	·	·	·	·	·	·	·	·
410	·	·	●	·	·	·	·	·	·	·	●	·	·	·	·	●	·	·	·	·	·	·	●	·	·	●	·	●
411	·	·	●	·	●	·	·	·	●	·	●	·	·	·	·	●	·	·	·	·	·	·	●	·	·	●	·	●
412	·	·	·	·	●	·	·	·	●	·	●	·	·	·	·	●	·	·	·	·	·	·	●	·	·	●	·	●
413	·	·	·	·	●	●	·	●	●	·	·	·	·	·	·	●	·	·	·	·	·	·	●	·	·	·	·	·
414	·	·	·	·	●	●	·	●	●	·	·	·	·	·	·	●	·	·	·	·	·	·	●	·	·	·	·	·
420	·	·	·	●	·	·	·	·	·	·	·	·	·	·	·	●	·	·	·	·	·	·	●	●	·	●	·	●
421	·	·	·	·	●	·	·	·	·	·	●	·	·	·	·	●	·	·	·	·	·	·	●	·	·	●	·	·
422	·	·	·	·	●	·	·	●	·	·	●	·	·	·	·	●	·	·	·	·	·	·	●	·	·	●	·	·
423	·	·	·	·	●	●	·	·	●	·	●	·	·	·	·	●	·	·	·	·	·	·	●	·	·	·	·	·
424	·	·	·	·	●	●	·	●	●	·	●	·	·	·	·	●	·	·	·	·	·	·	●	·	·	·	·	·
430	·	·	·	●	·	·	·	·	·	·	·	·	·	·	·	·	·	·	·	·	·	·	●	·	·	●	·	●
431	·	·	·	·	●	·	·	·	●	·	·	·	·	·	·	·	·	·	·	·	·	·	·	·	·	·	·	·
432	·	·	·	·	●	·	·	·	●	·	·	·	·	·	·	·	·	·	·	·	·	·	●	·	·	·	·	·
433	·	·	·	·	●	●	·	●	●	·	·	·	·	·	·	·	·	·	·	·	·	·	·	·	·	·	·	·
434	·	·	·	·	●	●	·	●	●	·	·	·	·	·	·	·	·	·	·	·	·	·	·	·	·	·	·	·
440	·	·	·	●	·	·	·	·	·	·	·	·	·	●	·	●	·	·	·	·	·	·	·	●	●	·	·	●
441	·	·	·	●	·	·	·	·	·	·	·	·	·	●	·	●	·	·	·	·	·	·	·	●	●	·	·	●
442	·	·	·	·	●	●	·	●	●	·	●	·	·	·	●	●	·	·	·	·	·	·	·	●	·	·	·	·
443	·	·	·	·	●	●	·	●	●	·	●	·	·	·	·	●	·	·	·	·	·	·	·	●	·	·	·	·
444	·	●	·	·	●	●	·	●	●	·	●	·	·	·	·	●	·	·	·	·	·	·	·	·	·	·	·	·

T Thermit, ES electroslag, OA oxyacetylene, B brazing, MMA manual metal arc, SA submerged-arc, EG electrogas, FC flux-cored arc, MIG gas metal arc, AST arc stud, TIG gas tungsten arc, ASP arc spot, PA plasma, PE percussion, FR friction, DF diffusion bonding, MB MIAB, F flash, RP resistance projection, RSE resistance seam, RSP resistance spot, RU resistance upset, PB power beam, US ultrasonic, EX explosive, S soldering, CP cold pressure, AB adhesive bonding. Processes to the right of FR are not generally applicable to structural fabrications or site welding. Processes towards the right of each half of each line can have less thermal affect on the workpiece.

List C

Principles and basic characteristics of welding and related processes

---

Adhesive bonding (AB)

Principle

A great variety of adhesives has been in use for many years but it is only in the last few decades with the development of synthetic resins that metals have been bonded reliably. These high strength adhesives are often known as 'engineering adhesives'. Adhesives must be fluid at the time the joint is made to allow the adhesive to flow and 'wet' each faying surface.

With clean metal surfaces free from moisture, oil, grease and loose foreign material an adhesive with a suitable viscosity will fill all the micro features of the metal surface. Subsequently the liquid adhesive is caused to harden by a chemical or thermal curing process, or with a hot melt adhesive by cooling. Hot melt adhesives soften with heat and are called thermoplastic. Cured adhesives are thermosets. Two part resin-hardener adhesives cure at room temperature or in a reduced time if heated. A large number of hardeners exist, allowing adhesive systems with a range of properties to be produced with a single adhesive. One-part adhesives need heat to activate the catalyst and initiate the cure. Anaerobic adhesives are unusual in that they are stable in the presence of air but once they are in a close fitting joint and oxygen is excluded they begin to cure.

The incorporation into an adhesive of particles of a resilient nature can change the behaviour in service. Plastisols are suspensions of PVC particles in a combination of liquids which on heating produce a semi-thermoset and elastomeric mass with high peel strength and an excellent gap filling capability. Epoxy adhesives have high strength in tension and shear but have poor peel strength and are not tough. This position can be improved by dispersing within the adhesive a rubbery phase which inhibits crack growth. Such adhesives are said to be 'toughened'. Electrically conducting adhesives are also available.

Method of use

As for soldering and brazing, some form of lapped joint is required with sheet. To prevent the joint peeling, particularly at the ends, additional spot welding or mechanical fastening may be used in these positions. This also holds the joint in contact while curing takes place. Spot welding can be carried out through the adhesive using a process called 'weld-bonding'. This is used in automobile manufacture to produce sealed joints with improved stiffness compared with solely spot welded joints. Surfaces should be clean and generally dry and free from grease, although some modern adhesives are tolerant of grease and oil. Anaerobic adhesives are used to lock studs and nuts and to fasten shafts into collars and gears. Here, adhesive strength can be adjusted to allow disassembly when required. Apart from these applications, adhesive bonding is used mainly on metal sheet and sections of less than 5mm thickness.

Dissimilar metals and metal to non-metal joints may be made. Joints to glass can be cured rapidly with UV light when using the appropriate adhesive. Compound structures such as honeycombs are readily made and stiffeners may be fastened to welded structures and panels in both original manufacture and for repair. Joints must be designed to put the bond into compression or shear and not to expose it to tension or peeling forces. A suitable choice of adhesive will give joints capable of withstanding temperatures up to about 250°C. Where joints must withstand moist conditions care in the selection of the adhesive and in preparation of the adherend surface, possibly with a special primer is necessary.

Adhesives are available in many forms, liquid, paste, mastic, solid, powder and tape and are dispensed with a range of tools e.g. rollers, brushes and both caulking and spray guns.

Advantages and limitations

Adhesive bonding has little or no effect on the properties of the parts being joined e.g. surface appearance, distortion and metallurgical properties. It is easily automated and joins and seals at the same time. Differing materials and thicknesses can be joined. Inaccessible joints are possible and it can join materials which cannot be welded.

Storage and handling of the adhesives requires special attention. Pre-treatment of parts may be necessary to avoid degradation in service and the joints have only a limited resistance to temperature. Joints must be restrained until curing is complete. Apart from the aircraft industry there is as yet no long term service experience for many applications.

Applications

The aircraft industry has a long and successful experience with adhesive bonding for both structural and secondary parts. Wide use is made of adhesives in the automobile industry for both car bodies and transmission parts. Other industries using the process are electronic, electrical and domestic equipment.

Factors affecting costs

Relative equipment cost, MMA=1. Varies widely according to the
degree of automation

Consumables – adhesives, hardeners, dispensing equipment

Mechanisation or automation

Can be mechanised

Skills required

Operator – none/moderate

Supervisor/maintenance engineer – moderate/high depending on
degree of automation

Further information

Lees W A (ed): 'Adhesives and the engineer'. Publ MEP Ltd, Bury St. Edmunds,
1989.

Arc spot (ASP)

Principle

There are two types of arc spot welding used on lap joints depending on
whether or not the electrode is consumable, one is based on MIG welding
the other on TIG. A timer is attached to the MIG or TIG equipment
allowing the process to be fired for a predetermined time and the gun or
torch is modified by the addition of a castellated gas nozzle. This allows the
welding gun to be pressed on the upper side of the overlapping parts to be
welded. A weld pool is created in this upper sheet which penetrates across
the interface and fuses with the lower member of the joint.

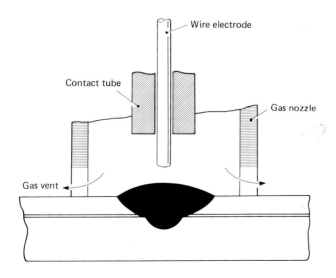

Arc spot welding using the
MIG process. When using TIG
the MIG gun is replaced by a
torch with a tungsten
electrode. The gas nozzle is
similar for both TIG and MIG
spot welding.

Method of use

The MIG version of the process is the more widely used and can be applied to both ferrous and non-ferrous joints. TIG spot is not generally suitable for aluminium alloys because of the difficulty in dispersing the interfacial oxide skin. Badly fitted joints should be avoided as only limited force can be applied to the welding gun to press parts together.

Advantages and limitations

A most useful process for joining overlapping joints in flat or vertical positions in which it is impossible to have access from both sides. There should not be too great a difference in the thickness of the parts if there is to be adequate fusion into the lower part. The upper part should always be the thinner of the two. A hole is not usually required in the upper member but may be provided to improve fusion into the lower part when welding thick material e.g. over 4mm, or for welding dissimilar compositions. Because of the difficulty of inspection it is not a process applied to critical applications.

Applications

Welding panels to stiffeners and attachments to structures in shipbuilding, automobile and general engineering.

Factors affecting costs

Relative equipment cost, MMA=1. The facility for MIG or TIG spot welding is often provided on MIG and TIG equipment as standard or can be obtained as a low cost extra.

Further information

Cary H B: 'Modern welding technology'. Publ Prentice Hall, Englewood Cliffs, USA, 1989, 2nd Edition, pp749–754.

Arc stud (AST)

Principle

An arc is initiated, usually by making contact between the end of the stud or similar attachment and the workpiece surface. After a predetermined time when a molten pool has been created in the workpiece and the end of the stud has fused, the stud is plunged into the molten pool. A ceramic ferrule surrounds the end of the stud to protect the arc from the atmosphere and to shape the molten metal which is forced out when the stud enters the molten pool. Although this ferrule gives adequate

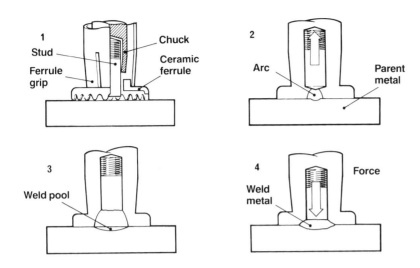

Principle of arc stud welding.

Typical arc stud welding equipment showing the power supply with controls to set time and current and the gun in welding position (courtesy Crompton Stud Welding).

protection for steel welds an additional stream of inert gas is required for aluminium welds. Steel studs have deoxidants on the stud end which melt into the molten pool and control weld quality. Depending on which version of the process is used the welding current may be cut off before the stud is returned to the surface or the current cut off may be delayed until the stud has actually entered the molten pool.

Method of use

Studs are gripped in a special hand held tool so that the operations described above can be carried out automatically. Many studs are cylindrical but rectangular cross sections can be used. Small studs may be welded in any position but larger studs can only be fastened in the flat position.

Advantages and limitations

Although the fastening of steel studs to steel plate is the most common application there is some scope for joining dissimilar metals. The fairly large molten pool limits the thickness on which the studs may be placed to not less than about 3mm. For placing studs on sheet thinner than 3mm the percussion (PE), or capacitor discharge method is preferred.

Applications

A major application is fastening shear connectors to beams and other surfaces in civil engineering for which purpose studs up to 20mm may be used. Other applications are fastenings for cover plates, insulation and panels and cleats for cables.

Factors affecting costs

Relative equipment cost (MMA=1) 4
Consumables – studs
Mechanisation or automation
  Semi-automatic, can be mechanised
Skills required
  Operator – none
  Supervisor/maintenance engineer – moderate

Further information

American Welding Society: 'Recommended practices for stud welding'. Publ AWS, (C 5.4-84) Miami, 1984, 34 pages.

Brazing (B)

Principle

A joining process different from welding in that under the action of heat a metal or alloy having a melting point lower than the parent metal is made to flow by capillary attraction into the space between the parts to be joined. A wide variety of brazing alloys are available and their choice depends on the parent metal and the service performance required. If the alloy used has a melting point lower than 450°C the technique is known as soldering.

It is necessary to have a gap of controlled dimensions between the parts to be joined to make use of the capillary effect and for the surface to be cleaned, often by the use of a flux which can be solid or gaseous. Invariably the joint design is a lap or some variation on this.

In both brazing and soldering there is no melting of the parent metal(s) but with a related technique known as 'braze welding' in which no reliance is placed on capillary attractions the joining is carried out using a technique like welding and the parent metal may be melted unintentionally.

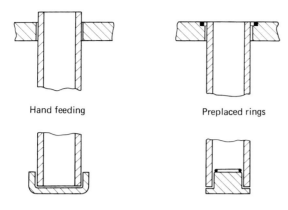

Hand feeding

Preplaced rings

Joint designs for hand feeding and preplacement of filler metal.

Method of use

The joint is designed to allow capillary attraction by having a gap sufficient for the flow of the appropriate flux and brazing alloy. Heat is supplied in a variety of ways. Both brazing metal in the form of rod and heat from a flame may be applied manually. Production rate can be speeded up, however, by preplacing the brazing alloy as a shim, ring, washer or powder and supplying heat by a flame (flame brazing) or by immersion in a bath of molten brazing alloy (dip brazing), by immersion in a bath of molten salts (salt bath or flux-dip brazing), by heat from HF induction (induction brazing) or a furnace (furnace brazing) or in a vacuum furnace (vacuum brazing).

Advantages and limitations

The method can be used on metals difficult to weld and on dissimilar metal and metal/non-metal combinations. Because of the uniform and diffuse heat employed distortion can be less than with a welding technique. The relatively prolonged heating of large areas does, however, soften those metals which have been work hardened or given low temperature heat treatments. The necessity for some form of lap joint can also be a limitation.

Applications

A very wide range is possible, from repairs, prototypes and short production runs using the simplest equipment to mass production of complicated assemblies such as heat exchangers and aircraft parts using furnace or vacuum brazing. Materials covered include carbon and alloy steels using copper based brazing alloys, high temperature and heat resisting steels using nickel based brazing alloys, non-ferrous alloys both copper and aluminium based and dissimilar metals including hard metal cutting tools. There are many possible filler metals and the correct choice is crucial to success. Advice can be obtained from the references below, from The British Association for Brazing and Soldering (BABS), or The Welding Institute.

Factors affecting costs

Relative equipment cost (MMA=1) 1–4 for manual equipment to 5 or 10
    for automated plant
Consumables – fuel gases and oxygen, fluxes and brazing alloys
Mechanisation or automation
    Can be mechanised
Skills required
    Operator – moderate
    Supervisor/maintenance engineer – moderate

References for further information

Brooker H R and Beatson E V: 'Industrial brazing'. 2nd edition
revised by P M Roberts, publ Newnes-Butterworth, London, 1975, 263 pages.

Cold pressure (CP)

Principle

Ductile metals such as aluminium and copper can be welded cold by
causing appreciable plastic flow at the interface so that oxide films on the
metal surface rupture and allow the freshly exposed metal to be brought
into intimate contact. The plastic flow is created in different ways
according to the type of joint but the aim is to extend the faying surface.
Surfaces to be joined must be clean and particularly free from water
vapour and oil films.

Method of use

Joints are either butt or a form of lap. Butt joints are used for welding
wires and rods. The freshly cut wire ends are pressed together in a pair of

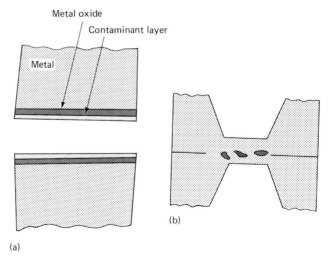

Mechanism of pressure
welding: a) Before welding,
surface oxide and
contaminants present; b) After
welding, oxide and
contaminants broken up and
dispersed.

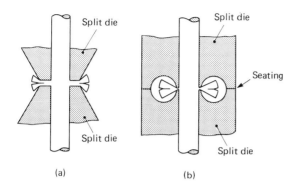

Dies for butt welding wire and rod: a) Flat faced; b) Self-trimming.

dies, which may be of the self-trimming type. Although spot welds may be made in overlapping sheets by forcing indenters into both sides such joints have little practical utility. Of greater use are the welds made in lapped or flanged metal in which the two thicknesses are cut through. This occurs in making longitudinal welds in cylinders where a flange is pinched off or in sealing a lid to a can where the press deforms and trims the flange. When copper is joined to aluminium the aluminium, being softer, must project further from the die to provide more flow.

Advantages and limitations

The process is rapid and has no thermal effect on the work. It is simple and cheap to operate once dies have been made but it is highly specialised as regards joint design and materials welded. Joints of the lap type must be considered early in design because of the need to allow for the deformation necessary. It is difficult to inspect cold welds and reliance must be placed on process control. Except for butt welds the thickness of the workpieces is reduced considerably at the weld.

Applications

The process is an excellent way of joining aluminium wires and rods and for making aluminium to copper connections for electrical purposes. The lap welded joints, apart from the tube closure type, have less utility.

Factors affecting costs

Relative equipment cost (MMA = 1) 0·3 – 5
Consumables – none
Mechanisation or automation
    Already mechanised
Skills required
    Operator – none
    Supervisor/maintenance engineer – moderate, process
    control important

References for further information

P T Houldcroft: 'Welding process technology'. Publ Cambridge University Press, 1977, 313 pages.
American Welding Society: 'Cold welding' pp 405 – 415 in Welding Handbook, 7th ed. Vol. 3, publ AWS, Miami, 1980.

Diffusion bonding (DF)

Principle

The parts to be joined are placed together and heated under modest pressure (5-15 N/mm^2) in vacuum or a controlled atmosphere. The temperature is sufficient to cause interfacial diffusion, 1000°C approx for steel. Surface oxide films dissolve in some metals to facilitate joining (steels and titanium) but in other metals e.g. aluminium which have refractory surface oxides an interfacial layer which provides a liquid film at bonding temperature to roll up the oxide film is necessary.

Method of use

Vacuum shielding is most commonly used and parts may be joined by being pressed inside a vacuum furnace. Large titanium structures have been made by sealing the parts in evacuated steel cans before squeezing in an open press.

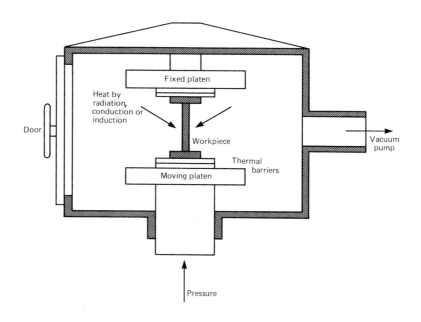

Vacuum enclosure for diffusion bonding.

Advantages and limitations

The process allows hollow shapes and other structures to be made which could not be fabricated in any other way. Dissimilar thicknesses and materials can be joined and distortion is negligible. The prolonged heating softens many metals, however, and diffusion bonded steels tend to have low impact properties. Inspection of the joint is difficult. Equipment is expensive.

Applications

Although many development trials have been carried out, some on large objects, there are relatively few production applications. These are either small items where dissimilar metals must be joined or larger structures in which, for example, water cooling passages or other internal spaces are required. Large titanium alloy structures have been produced for the aircraft industry.

Factors affecting costs

Relative equipment cost (MMA=1) 15+
Consumables – none
Mechanisation or automation
    Already mechanised
Skills required
    Operator – none
    Supervisor/maintenance engineer – moderate/high

Further information

The Welding Institute: 'Diffusion bonding as a production process'. Publ TWI, Abington, Cambridge, 1979.
Schwartz M M: 'Source book on innovative welding processes'.
Publ American Society for Metals, Metals Park, USA, 1981.

Electrogas (EG)

Principle

A mechanised vertical welding process in which an electric arc between a continuous solid or flux-cored electrode and the work piece fuses a large weld pool contained between water cooled shoes. The shoes are moved upward to keep pace with the welding. The process has many similarities with electroslag welding except that whereas in electroslag the heat is generated in a relatively deep flux bath by electric resistance, in electrogas there is only a thin layer of flux and heating is by an arc. The arc is shielded by a stream of gas, carbon dioxide, argon, helium or mixtures of these. The process is in effect a specialised form of vertical MIG welding.

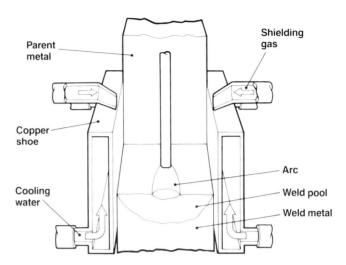

Principle of electrogas
welding.

Method of use

Edge preparation is generally a square edge gapped groove but can be a
single sided groove. Flux-cored wires are usually employed for welding
steels and bare wires with a suitable separate flux addition for non- ferrous
metals such as aluminium and titanium alloys. The shielding gases are
selected on the same basis as for MIG welding. Oscillation of the electrode
wire is preset but adjustable and vertical motion is automatic, usually
dependent on arc voltage.

Advantages and limitations

This is a fast method of automatic welding but it is limited to making joints
in the vertical or substantially vertical position. Good quality welds are
produced and because of the higher welding speed better HAZ properties
are obtained than with electroslag. Several types of machine exist, some of
them lightweight, which can be moved and positioned relatively easily over
the joint. Welding operators require adequate training.

Applications

Although the process has been used successfully on non-ferrous metals it is
normally used on structural steels, particularly in the manufacture of
storage tanks for which special types of equipment have been developed.

Factors affecting costs

  Relative equipment cost (MMA=1) 10 – 20
  Consumables – electrode wire and shielding gases
  Mechanisation or automation
    Already mechanised

Skills required
>   Operator – moderate/high
>   Supervisor/maintenance engineer – moderate/high

Further information

American Welding Society: 'Recommended practices for electrogas welding'. Publ
>   AWS, Miami, Fl, 1981, 48 pages.
American Society for Metals: 'Electrogas welding' in Metals Handbook, 9th Ed.
>   Vol. 6, publ ASM, Metals Park, Ohio, 1983.

Electroslag (ES)

Principle

This is a vertical method of welding in which a heated bath of slag is carried upward along the joint between water cooled dams. Heat is generated in the slag by the electrical resistance offered to a current passed between the workpiece and an electrode wire(s) fed into the slag bath. The wire melts and supplies filler metal to the joint. The water cooled dams, or shoes, are raised up the joint to keep pace with the addition of filler metal. Occasional additions of slag are made to the bath to replace that lost between the shoes and the work. Several forms of the process exist, wire electrode, consumable guide and plate electrode.

With the wire electrode one or more wires are introduced into the slag bath and are oscillated between the shoes to create even heating. The consumable guide method does not require the oscillating mechanism or that the wire guides are raised continually as the joint is made. A steel tube(s) which may be covered with flux is preplaced in the joint and a wire electrode(s) is passed inside the tube(s). Both the guide and the wire are fused into the joint. In the plate electrode process steel plates, one connected to each phase of a three phase supply, are lowered slowly into the slag bath.

Method of use

Square edge butt joints are employed with a gap of 30mm or more depending on the thickness to be welded. Both DC and AC power sources have been used but a DC constant voltage-constant wire feed system is now usual. A 3.2mm electrode wire is often used and may be solid or flux-cored. The powdered fluxes are similar to those used for submerged-arc welding but are higher in fluorides.

With the consumable guide method extra metal of special compositions may be incorporated into the consumable electrode to give weld metal of the desired analysis.

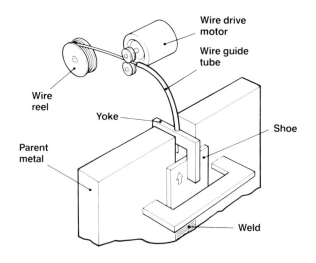

Principle of electroslag
welding.

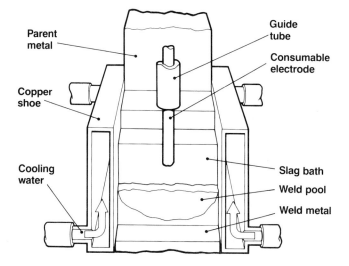

Wire electrode process.

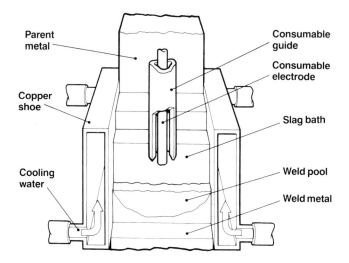

Consumable guide process.

Advantages and limitations

An economical process for joining heavy section mild and low alloy steels, which, although it can be used on metal as thin as 15mm, is not usually employed on plates thinner than 75mm. For this thickness the welding speed is about 1 m/hr. Only vertical or near vertical welds can be made. The prolonged heating and cooling cycle which the weld metal and HAZ undergo results in low notch toughness of the heat affected zone.

Applications

Vertical welds in vessels and heavy structural members in mild and low alloy steels where the low toughness can be accepted or where post-weld heat treatment can be applied. Variations of the process can be used for surfacing rolls and ingot refining.

Factors affecting costs

> Relative equipment cost (MMA=1) wire electrode 20,
> > consumable guide 4
> Consumables – electrode wire, guides, fluxes
> Mechanisation or automation
> > Already mechanised
> Skills required
> > Operator – moderate/high
> > Supervisor/maintenance engineer – moderate/high

Further information

Paton B E: 'Electroslag welding and surfacing', vol 1 & 2. English trans, MIR Publishers, Moscow, 1983, 256 and 264 pages.

Explosive (E)

Principle

A weld is produced when two pieces of metal are impacted at an appropriate angle and velocity, which should be less than the velocity of sound in the material being welded. The necessary velocity can be obtained by the detonation of an explosive charge. This explosive charge is placed against one part, usually the thinnest, which is known as the 'flyer'. The flyer is spaced away, by a distance known as the 'stand-off' from the second part which remains stationary and is placed on a block, often called the 'anvil. As the flyer folds down and spreads across the surface of the stationary part it behaves in a plastic manner and its surface and that of the stationary part is stripped off to form a jet which is ejected from the joint.

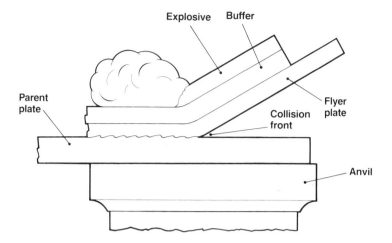

Explosive welding as used for cladding.

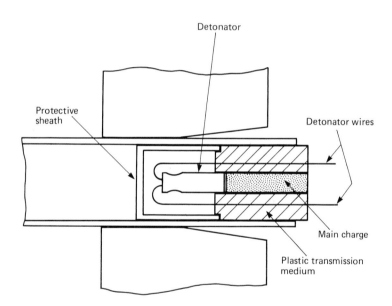

Explosive welding tube in a plate.

This jet takes with it the oxide and surface contaminants from the impacting surfaces and allows clean metal surfaces to be forced together to form the weld. The interface is rippled at roughly millimetre wavelength.

Method of use

When used for surfacing a sheet, explosive is placed on the flyer which is spaced away from the stationary plate. The required angle for welding develops as a result of the 'dog leg' which forms as the flyer folds down. Surfacing of thicknesses from fractions of a millimetre up to 5mm or more may be carried out using a wide variety of dissimilar metals. Tubular parts may be clad either internally or externally by placing the cladding material in tubular form, together with the explosive, inside or outside the part to be clad. In another form of the process, tubes are sealed into holes in tubeplates. In this application the tubeplate hole is tapered to provide the contact angle and the charge is within the tube. Another form allows the welding of socketted tube butt joints.

Advantages and limitations

The process can bond dissimilar metals, and thin sheet to thick plate. It is carried out at high speed with no thermal effect on the parts being welded. No source of electricity is needed. Special facilities are required, however, because of the security risk and noise.

Applications

Explosive welding is a highly specialised welding process requiring variations on the lap joint. The main uses are cladding down to foil gauges with dissimilar and special materials and tube plate welding.

Factors affecting costs

> Relative equipment cost (MMA = 1) Not applicable as only
> > consumables needed, however, a suitable area to carry out the process is necessary
> Consumables – explosive
> Mechanisation or automation
> > Not applicable
> Skills required
> > Operator – high
> > Supervisor/maintenance engineer – high

Further information

Crossland B: 'Explosive welding of metals and its application'. Publ Clarendon Press, 1982, Oxford, 250 pages.
Blazynski T Z (ed): 'Explosive welding, forming and compaction'. Publ London Applied Science, 1983, 402 pages

Flash (F)

Principle

Flash welding uses equipment comprising one fixed and one movable clamp so that the workpieces may be gripped and forced together. The work itself completes the single turn secondary of a heavy duty AC transformer. A voltage of 5 or slightly more is applied at the clamps and the parts are brought into contact. When current passes through the initial points of contact these are fused creating short-lived arcs and generating heat at the interface. Much of the molten metal at the interface is expelled as 'flash' and the moving platen is advanced to keep a constant gap until a suitable temperature has been reached at the interface. The parts are then

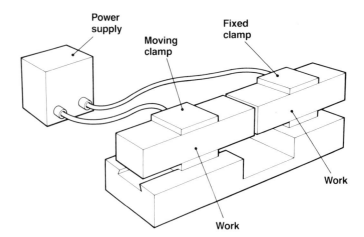

Principle of flash welding.

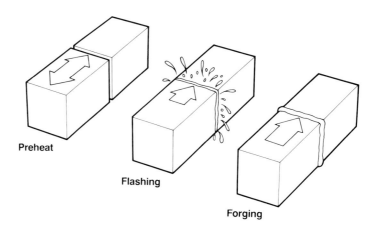

Flash welding in three stages.

Typical equipment (courtesy Verson AI).

forged together. When welding heavy parts preheating may be employed by bringing the parts into contact a number of times and then separating them to create resistance heating but delaying the onset of flashing.

Method of use

The process is particularly suited for joining bar stock and compact sections to themselves or to matching forgings, castings or machined shapes but with suitable plant it can be used to join lengths of sheet or plate. The contacting parts should be of roughly similar shape. If one is a flat surface e.g. as in joints 15, 20 and 40, a projecting rib must be raised in the surface to provide the matching contact. Mild, carbon-manganese and alloy steels are extensively welded by flash welding and with care in choice of equipment non-ferrous metals including aluminium alloys may be welded. The size range of applications is considerable. Efficient clamping of parts is essential.

Advantages and limitations

The process is a fast, mechanised method of joining which, because most fused metal is squeezed from the joint, gives high quality welds in steels and dissimilar metal combinations. The flash left on the joint is irregular and is usually removed by special stripping dies in the welding machine or later by machining or grinding. Hardenable steels may require post-weld heat treatment. Large machines place an uneven and intermittent demand on the mains and the flash ejected is unpleasant and messy.

Applications

Widely used throughout industry for joining sections and bars and for attaching fittings to rods and sections. Used for automobile transmission gear and wheel rims, aircraft engine rings, window frames and steel rails.

Factors affecting costs

Relative equipment cost (MMA=1) 5 to 100 or more depending on
    capacity
Consumables – none
Mechanisation or automation
    Already mechanised and can be fully automated
Skills required
    Operator – none
    Supervisor/maintenance engineer – moderate, depends on degree
    of automation

Further information

American Welding Society: 'Welding handbook', Vol. 3. Publ AWS, Miami, Fl.

Flux–cored arc (FC)

Principle

An arc welding process using a continuous hollow electrode filled with a flux which provides shielding gases, deoxidisers, alloy additions and slag formers. Some cored electrodes are designed to be used with an additional gas shield which is usually carbon dioxide or a carbon dioxide rich gas mixture. An argon based gas shield is always used with the metal cored wires, in which the core is largely powdered iron and alloying elements. When there are enough materials in the core to provide shielding and deoxidation no external shielding gas is necessary and the wires are said to be 'self-shielding'. Such wires allow welding which is reasonably resistant to draughts. The equipment used is basically the same as for MIG welding but the generally higher currents which can be used with cored wires require higher capacity power sources.

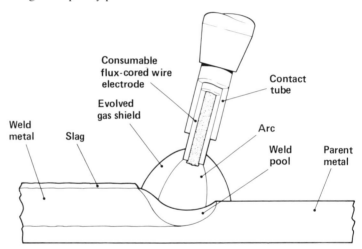

Principle of flux-cored arc welding.

Method of use

Both manual and mechanised welding are employed, using a DC power supply, DCEP for rutile cored and hard facing wires but DCEN can be used for basic and metal cored wires. Wire sizes from 1.2mm to 2.4mm diameter are generally employed but wires up to 4mm diameter are often used for surfacing. Because the current is carried almost entirely by the metal sheath the wires are particularly susceptible to resistance (I^2R) heating which gives higher burn-off and deposition rates than for the same diameter of solid wire. Rutile wires are easy to use and give smooth flat standing fillets. For improved low temperature properties and when welding thick steel a 'basic' wire is needed. Higher strength and improved low temperature properties require wires containing nickel and molybdenum. Wires for stainless steel are frequently made with low carbon steel sheaths and have the alloying elements in the core but with small diameter wires some manufacturers employ a stainless steel sheath. Wires for hard facing have the alloying elements wholly within the core. With self-shielding wires the welding gun can be simplified as a gas nozzle is not required and the lack of sensitivity to air allows improved access to difficult joints by permitting a long electrode extension, even up to 35mm.

Advantages and limitations

Flux-cored wires have high deposition rates and give deep penetration. Their use in preference to solid wires is therefore generally taken on economic grounds. The presence of a flux allows positional welding where it would not be possible with solid wires. Weld quality is good and the weld metal has a low hydrogen potential. Self-shielding wires are more resistant to draughts than metal arc electrodes. Flux-cored wires may give more fume than some metal arc electrodes.

Applications

The process is widely used in structural engineering, earth moving plant, shipyards and offshore fabrications. It is frequently used for welding stainless steels and widely employed for hard surfacing. The equipment can be operated by a robot.

Factors affecting costs

Relative equipment cost (MMA=1) 2 – 3
Consumables – electrode wires, shielding gases if required
Mechanisation or automation
    Semi-automatic, can be mechanised
Skills required
    Operator – moderate
    Supervisor/maintenance engineer – moderate

Further information

Hobart Brothers Company: 'Technical guide for flux cored arc welding'. Publ Hobart Brothers, Troy, Ohio, USA, 1981, 106 pages.
Houldcroft P T and John R: 'Welding and cutting'. Publ Woodhead-Faulkner, Cambridge, pp 102–116.

Friction (FR)

Principle

This process uses the generation of heat by friction to join mainly tubular or solid bar parts. The parts to be joined are pressed together and one is rotated so creating frictional heat at the interface which is rendered plastic. Rotation is then stopped and a forge is applied to consolidate the joint as it cools. Two forms of this basic process exist, *continuous drive* in which the

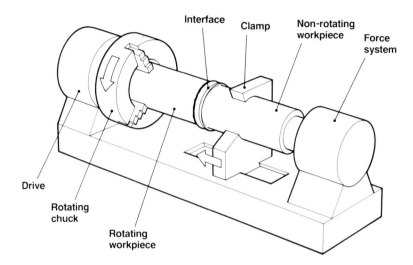

The basic friction welding process.

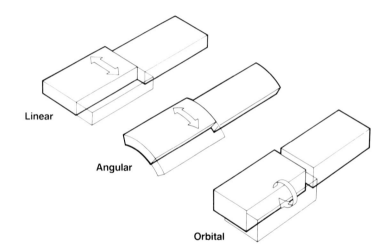

Alternative motions.

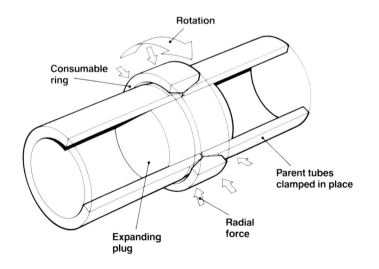

Radial friction welding.

motor is always connected to the work and starts and stops with each welding operation and *inertia welding* in which the motor drives a flywheel which is disconnected from the motor to make the weld. Plastic metal is extruded from the interface along with oxides and surface impurities forming a collar around the joint. No fusion is involved and the joints are of high quality. Dissimilar metals may be joined e.g. stainless steel and aluminium and some plastics. More recent developments of the process permit linear or orbital motion allowing the joining of rectangular shapes. In another form *radial friction welding* a ring is revolved and compressed on to the rigidly clamped workpieces. Radial friction welding can be used to join tubes or to put rings on tubular or solid bars.

Method of use

Friction welding is an extremely flexible and tolerant process. It has been employed to weld bar from 1mm diameter up to over 100mm or the equivalent area in tube form. Rotational speeds and machine capacities vary from 80 000 rpm and a few kg load for 1mm to 40 rpm and hundreds of tonnes load for the largest sizes. Although machines may be loaded and unloaded manually automatic machines are in wide use. Weld times are from 1 to 250 sec over the range mentioned.

Advantages and limitations

Friction welding is a high quality, fast, completely automatic method of joining steels and dissimilar metals. It is limited to joints in which at least one member is of circular or tubular shape unless one of the recently developed variations is used. Compared with flash welding it is a cleaner process which places a balanced and steady load on the mains.

Applications

An ideal process for attaching forgings or other items to shafts or bars. Double ended machines exist e.g. for joining hub bearings to axles. Widely used in the automobile component, engine and agricultural machinery industries and for welding high speed steel ends to twist drills.

Factors affecting costs

Relative equipment cost (MMA=1) 5 – 150 depending on capacity
Consumables – none
Mechanisation or automation
    Already mechanised
Skills required
    Operator – none
    Supervisor/maintenance engineer – moderate

References for further information

Schwartz M M (ed): 'Source book on innovative welding processes', pp 1–75. Publ American Society for Metals, Metals Park, USA, 1981.
The Welding Institute: 'Exploiting friction welding in production'. Publ TWI, Abington, 1979, 80 pages.

Manual metal arc (MMA)

Principle

An arc between the workpiece and an electrode comprising a metal rod (core wire) covered with flux provides heat and extra metal to fuse and fill the joint. A suitable choice of metal and flux covering allows the process to be used for a wide variety of applications, different metals and welding positions. The core wire melts away readily leaving a cup shaped end to the electrode, the depth of which depends on the flux covering. Metal is transferred from the core wire to the molten pool in the workpiece as either a stream of small drops or as large globules which touch the molten

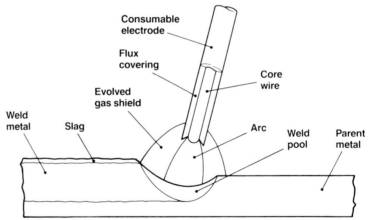

Principle of manual metal arc welding.

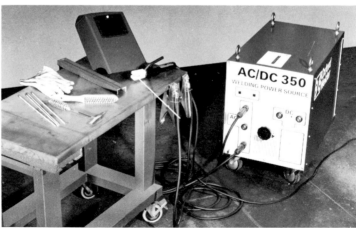

Equipment for welding with a transformer/rectifier power plant (courtesy Oxford Products).

pool. The flux covering provides protection to the molten pool and the transferring drops by providing a film of molten flux and gases such as carbon dioxide. It also provides alloying elements where necessary to give the weld metal an appropriate analysis. A residue of fused flux covers the completed weld and is removed by chipping and brushing. Because the electrodes have a finite length the MMA process is intermittent with the molten pool solidifying while the remains of one electrode are removed from the electrode holder and the next is fitted.

An important characteristic of the flux covering is its basicity which is determined by the amount it contains of such compounds as calcium carbonate and calcium fluoride. Basic electrodes, high in these compounds, give tough high quality weld metal and can be made to provide the low hydrogen deposits necessary to avoid cracking in higher carbon and alloy steels. Another important constituent of coverings is the mineral rutile which gives a smooth running and easily used electrode but one giving welds lacking the toughness and low hydrogen characteristics of welds made by basic electrodes. Iron powder may be added to coverings of all types to give smoother operation with less spatter and to increase the rate of metal deposition.

Method of use

Electrodes have core wires ranging from 1.6mm to 6.3mm (occasionally 10mm) with lengths from 250-450mm. The electrode diameter used depends on the thickness of metal being welded. Small diameters are used on thin sheet or in the root of grooves forming the edge preparations in thick material where good access to the root is important. Smaller electrodes are used when the welding is carried out in vertical and overhead positions than in the flat position. The electrode holder is held by the welder who must feed the electrode towards the molten pool keeping a constant arc length and at the same time also moving along the joint. When making wide or multi-pass welds it is often necessary for the welder to manipulate the electrode so that the arc moves in a pattern. Care is needed to fuse correctly the crater left in the weld bead when the previous electrode was used up. Except for the 'gravity' welding variation it is a manual process. In gravity welding the electrode which is longer than for manual welding is held in a simple inclined mechanism down which it slides as it burns away.

Advantages and limitations

This is the most important general purpose welding and surfacing method using low capital cost equipment. With an appropriate choice of electrode it is capable of welding many metals and alloys up to high quality standards. The electrode and its holder are small and can be used at the end of a long cable and in inaccessible positions. The power source is simple and is readily transportable. Mains operated AC transformers and transformer-rectifier sets are common but petrol or diesel driven equipment is available for site welding. Because it is an intermittent process and manually operated it has been replaced in some applications by MIG or flux-cored wire welding which are continuous processes giving higher productivity. Weld quality is determined mainly by operator skill.

Applications

A wide range of metals and applications can be welded, structural steelwork, shipbuilding, general engineering, process plant, pipework repairs and maintenance. Both shop and site welding are carried out in sheet metal and thick plate. Gravity welding is used mainly for welding stiffeners to panels in shipbuilding and structural work.

Factors affecting costs

Relative equipment cost (MMA=1)
    1 for mains operated power source
    2 for portable engine driven power source
Consumables – electrodes
Mechanisation or automation
    Manual, only mechanised form is gravity welding
Skills required
    Operator – moderate/high
    Supervisor/maintenance engineer – moderate, depending on quality required

Further information

Houldcroft P and John R: 'Welding and cutting'. Publ Woodhead Faulkner Ltd, Cambridge, 1988, 232 pages.
Davies A C: 'The science and practice of welding'. Publ CUP, 9th ed, Vol 2, Cambridge, 1989.

Metal inert gas or gas metal arc (MIG)

Principle

An arc welding process employing a continuous solid electrode and a gas to shield the arc and molten pool. Depending on the metal being joined and the type of metal transfer desired from the electrode the shielding gas may be inert (argon or helium), carbon dioxide or mixtures of these with or without oxygen. For manual or robot operation the electrode is of small diameter, 0.8mm to 1.6mm, fed at a constant speed. A consistent arc length is maintained by what is called the self-adjusting effect in which the constant voltage (flat) power source causes current changes which oppose arc length changes. Flexibility of operation is provided by separating the welding gun from the wire feed unit with a flexible conduit down which the wire is fed. For heavier duty mechanised operation, when wires up to 2.4mm diameter may be used, the wire feed unit is fitted directly over the

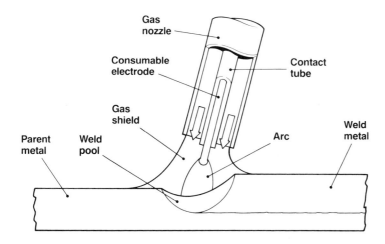

Principle of MIG welding.

Equipment for MIG welding
with goose-necked gun
(courtesy Oxford Products).

gun. Although the self-adjusting arc can still be used with these thicker
wires it is sometimes preferred to control the arc length by varying the wire
feed speed in response to a voltage signal.

Method of use

MIG welding is carried out on DC with the electrode positive (DCEP). A
distinctive feature of MIG welding is the way the shielding gas and welding
conditions can be selected to give the desired form of metal transfer from
the electrode. At currents over about 200A metal is transferred as a spray
allowing high deposition rates but only in flat and HV positions. Inert gases
are used for non-ferrous metals and inert gases with oxygen or carbon
dioxide additions for steels. With steel wires the voltage can be reduced
markedly and then lower currents give a form of transfer called 'dip'. Metal
is fused directly into the molten pool and with this condition welding can be
carried out on thin metal and in vertical and other positions. Dip transfer
welding is done in an atmosphere of carbon dioxide or argon-CO_2 mixtures

and it only works with higher resistivity wires like steel. Aluminium wires cannot be used with this technique. To use MIG welding on light gauge metals of all types including aluminium and stainless steel a 'pulsed' arc using a special power source may be necessary.

Advantages and limitations

With the appropriate choice of shielding gas and welding conditions this process is extremely versatile allowing the welding of most ferrous and non-ferrous alloys in a wide range of thicknesses and all welding positions. High quality weld metal is produced and the completed welds are clean and do not usually require deslagging although small islands of glassy slag may occur on the surface when welding steel. The working area must be protected from draughts and the equipment is more expensive to buy and complicated to maintain than that for manual metal arc. It is also less portable. Being a continuous process it is suitable for mechanisation and operation by welding robots. The equipment is basically the same as that required for flux-cored arc welding.

Applications

More weld metal is now deposited by the MIG process and its related cored wire process than MMA which for fifty years was the most widely used process. Unlike the other inert gas shielded process, TIG welding, the MIG process is ideally suited for the making of fillet welds. It is indispensable for welding non-ferrous metals such as aluminium and copper based alloys and in both manual and automatic versions is used throughout industry e.g. shipbuilding, structural, process plant, electrical, domestic equipment and automobile industries.

Factors affecting costs

Relative equipment cost (MMA=1) 3 to 10 depending on the output and complexity of the power source

Consumables – electrode wires, shielding gases, replacement guide tubes and nozzles

Mechanisation or automation
Semi-automatic, can be mechanised

Skills required
Operator – moderate
Supervisor/maintenance engineer – moderate/high depending on degree of automation

Further information

Houldcroft P T and John R: 'Welding and cutting'. Publ Woodhead-Faulkner, Cambridge, pp 75–101.
Hobart Brothers Company: 'Technical guide for gas metal arc welding'. Publ Hobart Brothers, Troy, Ohio, USA, 1980, 106 pages

MIAB (MB)

Principle

MIAB stands for *magnetic impelled arc butt*, a welding process for joining thin walled tubes and hollow sections. The parts to be joined are clamped into the machine, one clamp of which is movable. Magnetising coils surround each of the parts to be joined and the coils are split to enable them to be placed around the work. The parts are moved together and then separated to create an arc. Under the action of the magnetic field provided by the coils this arc moves rapidly around the weld line melting the faces of the joint. After 0.5 to 10 seconds depending on the size and thickness of the work, the joint faces are fused sufficiently and the parts can be pushed together to consolidate the joint and squeeze out oxidised metal to form a flash around it.

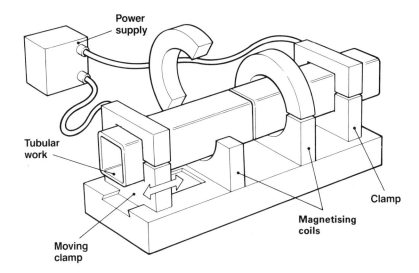

Equipment for MIAB welding.

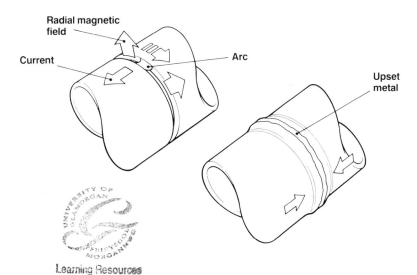

MIAB welding in detail.

Method of use

MIAB welding is usually employed for butt welding thin walled tubes or hollow sections to each other or to pressings, forgings or machined parts with matching contours. Shielding is not necessary for welds in steels but may be used to improve the appearance of the metal extruded from the joint when it is upset. Gas shielding may also be used on some non- ferrous parts but not aluminium alloys. Although special plant allows thick tubes to be welded the process is normally limited to a maximum of 5mm thickness.

Advantages and limitations

The process is particularly suitable for joints in steel which are less than 3mm thickness and 200mm circumference. Non-magnetic materials require the insertion inside the tubes or shape of a ferrous plug to concentrate the lines of magnetic force. The process can be applied to non-circular tubes. Compared with flash welding it is cleaner, quicker and uses lower upset forces. It is also quicker than friction welding and can weld thicknesses too thin to be gripped in a friction welder and because there is no rotation there is no location problem with non-symmetrical parts.

Applications

Although there is a wide potential for the process it is at present used chiefly for the mass production of automobile components.

Factors affecting costs

Relative equipment cost (MMA=1) 5 – 10
Consumables – usually none but shielding gas may be used
Mechanisation or automation
   Already mechanised
Skills required
   Operator – none
   Supervisor/maintenance engineer – moderate

Further information

Johnson K I et al: 'MIAB welding, principles of the process'.
Metal Construction 1979 **11** (11) 590–597.

Oxyacetylene (OA)

Principle

Equal volumes of oxygen and acetylene at the same pressure are fed from cylinders to a blowpipe where they are burnt to produce a flame. A neutral

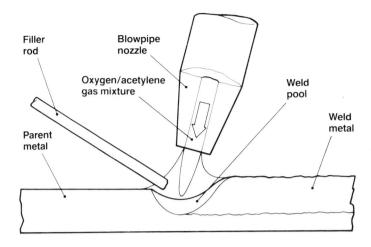

Principle of oxyacetylene welding.

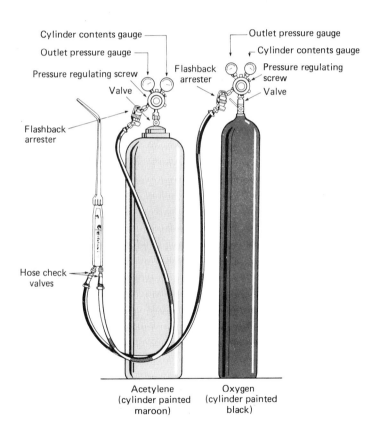

High pressure gas welding equipment.

flame is used for most welding. An excess of acetylene results in free carbon which makes the central part of the flame white and luminous. This is called a carburising flame and while unsuitable for welding it can be used for surfacing steel. Oxidising flames resulting from an excess of oxygen are sometimes used for welding metals containing a high proportion of zinc where the effect is to suppress to some extent the vaporisation of the zinc. When welding mild and carbon-manganese steels the CO and H in the flame both clean and protect the surface from oxidation. With other metals such as stainless steel, copper alloys and aluminium and its alloys a flux is necessary to perform these functions.

Method of use

Oxyacetylene (or gas) welding is invariably carried out manually. Compared with the electric arc it is a diffuse heat source which allows excellent control of the welding process. This permits unsupported welds to be made such as the root runs in pipes and the bridging of gaps in edge preparations. A skilled welder can make welds in all welding positions.

Advantages and limitations

The equipment is simple, relatively cheap and portable. Excellent control of welding is possible but the diffuse nature of the heat source can cause serious softening and distortion of the workpiece.

Applications

For butt joints in sheet metal and root runs in thicker prepared joints.
    Widely used for repair welding.

Factors affecting costs

    Relative equipment cost (MMA = 1) 0.5
    Consumables – electrode wires, gases and fluxes
    Mechanisation or automation
        Manual, cannot be mechanised
    Skills required
        Operator – moderate
        Supervisor/maintenance engineer – minimal

Further information

Houldcroft P and John R: 'Welding and cutting'. Publ Woodhead Faulkner Ltd, Cambridge, 1988, 232 pages.
Parkin N and Flood C R: 'Welding craft practice'. Vol 1, 'Oxy-acetylene gas welding and related studies', 2nd. ed. Publ Pergamon Press, Oxford, 1979, 161 pages.

Percussion (PE)

Principle

Percussion welding appears in several forms but in each a high intensity arc of short duration fuses the surfaces of the work pieces before they are impacted 'percussively' to consolidate the weld. The energy for the arc is

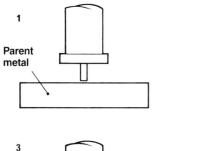

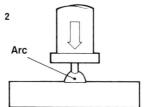

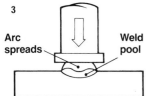

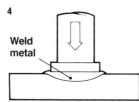

Stages in making a percussion weld.

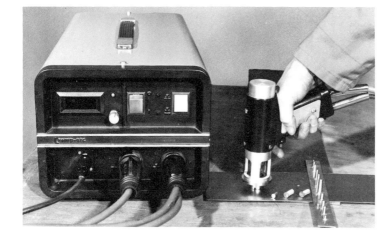

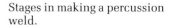

Equipment for percussion welding (courtesy Crompton Stud Welding).

usually stored in a capacitor hence the alternative name for the process – capacitor discharge welding. The main difference between the various forms of percussion welding is in the way the arc is ignited. In a widely used form employed for stud welding a small tip is machined on the end of the stud. This is held in contact with the work in a spring loaded device. When the charged capacitor is connected this tip is vaporised allowing an arc to form and the stud to be impacted on to the work under the action of the spring.

Method of use

Percussion welding is mainly used for attaching wires to surfaces or studs to thin sheet. When welding wires the capacitor is discharged by dabbing the wire on the work. With stud welding the spring loaded holder for the stud is often in the form of a portable handgun. Because of the short duration of the arc the depth of fusion is extremely low, perhaps even 0.1mm. This enables studs to be welded to thin sheet without damaging the opposite surface which is often painted, coated or polished. The thin fused layer also allows dissimilar metals to be joined e.g. brass studs to steel sheet or steel studs to zinc base die castings.

Advantages and limitations

The process is a rapid method for attaching studs with minimum heat effect on the workpiece. It can be applied to studs between 3 and 8mm diameter using portable equipment. A range of dissimilar metals may be joined.

Applications

Studs, fastenings and other small fittings are attached to automobile, domestic, electrical and architectural products. A variation of the process is used to weld electrical contacts on to their supports.

Factors affecting costs

> Relative equipment cost (MMA = 1) 1 – 4
> Consumables – only the studs or other attachments
> Mechanisation or automation
> > Can be mechanised
> Skills required
> > Operator – none
> > Supervisor/maintenance engineer – moderate

Further information

American Welding Society: 'Welding handbook', Vol 3. Publ AWS, Miami, Fl.
Houldcroft P T: 'Welding process technology'. Publ CUP, Cambridge, 1977.

Plasma (PA)

Principle

Plasma arc is a development of TIG in which the arc is forced through a water cooled collar or orifice. This concentration of the plasma of the arc raises its temperature and concentrates it into a narrow columnar shape which is highly directional and capable of penetrating deeply into the workpiece. DCEN or square wave AC is employed. Unlike TIG torches the plasma torch allows two separate flows of shielding gas to be used, argon is normally preferred for the flow through the orifice (the plasma flow) but gas mixtures can be used for the shielding flow. The process exists in four forms: (1) Micro-plasma, 0.1–20A; (2) Medium current, < 100A; (3) Keyhole plasma, > 100A; (4) Cutting.

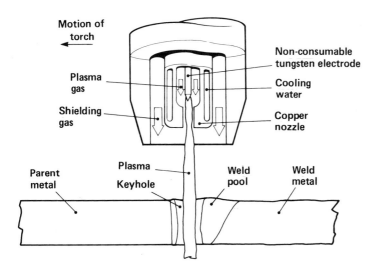

Principle of plasma welding.

Method of use

Micro-plasma is often preferred to low current TIG as it gives a needle-like stable arc which is more easily directed than the low current TIG arc. Medium current plasma competes with conventional TIG over which it has the advantage of easier arc striking, less chance of tungsten contamination and smoother more deeply penetrating welds. It is often used in mechanised welding. In keyhole plasma welding the penetration is deep and narrow and allows workpieces up to 6mm thickness to be penetrated completely (see under power beam welding for an explanation of keyholing). This allows welds with smooth underbeads to be formed in joints where access cannot be obtained to the underside. The cutting mode is similar to keyholing but with higher currents and gas flows. Argon-hydrogen mixtures are used for the plasma gas when cutting.

Advantages and limitations

Micro-plasma can be used for delicate work such as thin sheet, wire and mesh for which conventional TIG would not be feasible. In the medium current range the easier starting and superior weld finish make the process particularly suitable for mechanised welding. Both medium current and keyhole plasma, however, require plant more costly to buy and maintain than for TIG welding. Plasma cutting is a most useful process for cutting non-ferrous metals not readily cut by oxyacetylene, while a recent development, air-plasma, has great potential for high speed cutting of sheet steel.

Applications

Plasma welding is a precision process which can produce high quality welds in a range of ferrous and non-ferrous metals. It can successfully weld thin metal and forms such as mesh which cannot be welded by any other process except power beam. The process is mainly used in aircraft engine, aerospace, chemical plant and automobile equipment industries.

Factors affecting costs

Relative equipment cost (MMA=1) 4 – 8
Consumables – plasma and shielding gases
Mechanisation or automation
   Readily mechanised
Skills required
   Operator – moderate
   Supervisor/maintenance engineer – moderate

Further information

Lucas W: 'TIG and plasma welding'. Publ Abington Publishing, Cambridge, 1990.

Power beam (PB)

Principle

The electron and laser beam welding processes have become known collectively as 'power beam' welding. Although the two processes are very different in the way the beams are generated and in the nature of the beams themselves the type of weld produced is similar. For welds up to about 5mm thickness in steel the processes are usually interchangeable but above this thickness electron beam with its greater efficiency and higher available power becomes progressively preferable.

In the electron beam process a stream of electrons is generated in an electron gun and focused electro-magnetically on the work. The gun and work must generally be within a vacuum enclosure. At the work the electron beam can have such intensity, over $10\,\mathrm{kW/mm^2}$, that it vaporises a cavity through the complete thickness of the joint. This process is known as 'keyholing' and it allows deep narrow welds with minimum heat affected zones. As electron beam welding is carried out in a vacuum no shielding is necessary.

The radiation from a laser can be transmitted through air and is focused optically on the workpiece to similar energy densities to electron beam. Keyhole welds are also made with the laser. As the operation is carried out in air the weld must be shielded with helium or argon.

Method of use

Electron beam equipment can have electron guns with a capacity of <1kW to 100kW and a corresponding capability for welding thicknesses from fractions of a millimetre to 200mm. The size of the work which can be handled depends on the capacity of the vacuum chamber, many of which are designed for specific applications. Laser beam equipment is of two main types, solid state, mainly pulsed equipment up to 500W, and

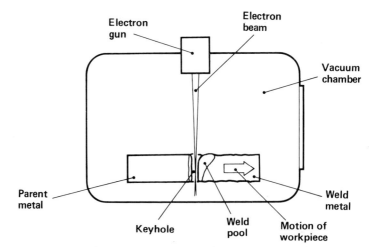

Principle of electron beam welding.

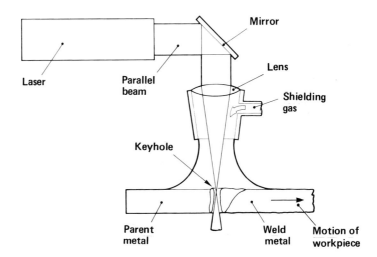

Principle of laser beam welding.

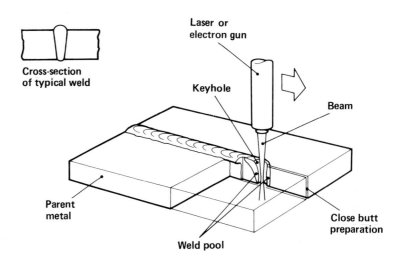

Mechanism of keyhole penetration.

continuous CO_2 lasers from 250W to 25kW, although above 5kW the cost of the equipment escalates. Lasers have taken over many of the applications up to 500W particularly in the micro-electronic field and are increasingly used up to 5kW in preference to electron beam for making welds up to 5mm deep. A major reason for this development is that, not requiring a vacuum chamber, the production engineering and work handling is simplified.

Solid state lasers can produce spot welds but may also produce continuous welds by stitch welding. They are also used for drilling fine holes. With defocused beams and with a raster beam manipulation the continuous lasers can be used for surface heat treatment and weld surfacing.

Advantages and limitations

Both electron and laser beam welding being high intensity heat sources allow welds to be made with a minimum spread of heat in the work. Very high welding speeds can be obtained with a minimum of distortion. The keyholing technique allows deeply penetrating narrow welds with little rotational distortion so that it is often possible to make welds in finish machined parts. Accurate machining of parts is usually necessary and self-locating joints are desirable. The simplest parts to weld are those with a circular joint line as they can be revolved under the beam. Long straight joints must be set up accurately because of the narrow weld profile but tracking devices are feasible for both electron beam and laser. The equipment is expensive but productivity is high and weld quality excellent. Often the use of power beams allows products to be made which could not be fabricated in any other way.

Applications

Pulsed laser and small electron beam equipment is used for welding electronic components and small scale welding. Electron beam and increasingly laser beam are used for automobile components e.g. gears and transmission assemblies, for aircraft engine and aerospace products, also domestic appliances. Higher capacity electron beam equipment is used for pressure vessel, nuclear, process plant and chemical plant. The power beam processes are used particularly for high volume production or for critical applications. Because of equipment costs much use is made of jobbing shops.

Factors affecting costs

Relative equipment cost (MMA=1)
Small equipment 15 – 40
General purpose 50 – 200
High power and special purpose 100 – 750
The high cost of plant is in part a result of the requirement for precise high speed work manipulation equipment, control and monitoring systems

Consumables – electron beam, none; laser, lasing gases and shielding gases

Mechanisation or automation
   Already mechanised

Skills required
   Operator – Low/moderate, depending on whether or not job is part of an established production run
   Supervisor/maintenance engineer – moderate/high

Further information

Schwartz M M (ed): 'Source book on electron beam and laser welding'. Publ American Society for Metals, Metals Park, USA, 1981, 398 pages.
The Welding Institute: 'Power beam technology'. Proc int conf, The Welding Institute, Cambridge, 1987, 446 pages.

Resistance projection (RP)

Principle

A process in which the heat for welding is generated by the resistance offered to the passage of an electric current through limited points of contact between the parts to be joined. The points of contact are created by pressing or machining projections on one part or by the natural contours of the parts when brought together. During welding the heat generated in the projection and the place where it contacts the other part of the workpiece causes the projection to collapse and the two parts to butt together under the pressure of the welding head. This collapse is called 'set–down' and the component to be welded must be designed to allow this to happen. The operation is carried out in a press welder with the work between copper platens which are in the single turn secondary of a heavy duty transformer. Some machines have a facility for sloping–in and

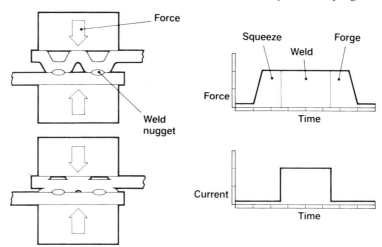

Principle of resistance projection welding.

sloping–out the welding current which increases the tolerance of the process and reduces splashing of hot metal from the joint.

Method of use

Projection welds are used singly or in groups of two, three or four made simultaneously and not as with spot and seam welds in an array of many welds to be made in sequence to form a continuous joint. The only continuous projection weld is that when an annular projection is used to close an opening. Alignment and other difficulties limit the size of annular projections to a seam no longer than about 200mm. It is unusual to employ the process on any material other than low carbon mild steel.

Advantages and limitations

Projection welding is an ideal method of fastening attachments e.g. brackets, spigots and weld nuts to sheet metal where there is access from only one side and for making attachments to solid forged or machined parts. Short length T joints e.g. 14 or 15 can be made by forming projections in the edge of the stem. Crossed wire welds are also possible in a variety of materials including aluminium alloys. Welds can be made at high speed with limited heat affect on the parts. Alloy and hardenable steels may require a subsequent heat treatment to produce ductile joints. No consumables are used in projection welding. Once the machine has been set little operator skill is required.

Applications

The process is widely used in automobile body and component manufacture and in domestic and electrical equipment.

Factors affecting costs

Relative equipment cost (MMA = 1) 2 – 10
Consumables – none
Mechanisation or automation
    Already mechanised, may be hand fed or fully automated
Skills required
    Operator – none
    Supervisor/maintenance engineer – moderate

Further information

American Welding Society: 'Welding handbook', Vol 3. Publ AWS, Miami, Fl., 1980, 7th. ed.

Resistance seam (RSE)

Principle

A resistance welding process in which overlapping sheets are gripped between roller electrodes enabling a series of either discontinuous or overlapping spots to be made. The force on the electrodes is sufficient to give good contact at the surface of the sheets and a fused nugget develops at the faying surfaces. Welding current is usually supplied in pulses with rotation being continuous but some use is also made of continuous current. As the previously made nugget provides a shunt path for the welding current a higher current is needed than for spot welding (RSP) of the same thickness material.

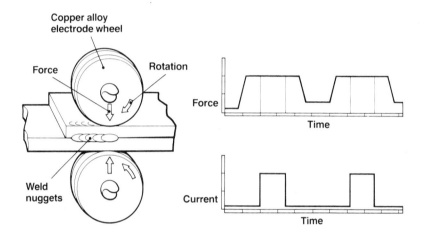

Principle of resistance seam welding.

Method of use

Resistance seam welding is used for making drums when it is employed for both the longitudinal seam and flanged ends. A variation in the method allows the overlap in sheets to be forged down during welding to produce a flush joint. This is called 'mash–seam' in which the sheet overlap is no more than 1½ times the material thickness. Seam welding is used on material from 0.5mm to 3mm thickness. It is applied to mild and alloy steels, stainless and heat resisting alloys, aluminium and copper alloys.

Advantages and limitations

Resistance seam welding is a comparatively rapid method of welding which is mechanised and readily controlled as regards quality. The joint is, however, a lap which is unacceptable for applications in which the notch would provide a seat for the start of corrosion or fatigue. The equipment is more expensive to maintain than other resistance welding plant. It is unusual to use the process for welding metal thicker than 3mm because of the mechanical difficulties in applying pressure and the wear on the electrode wheels.

Applications

Drums, cans, aircraft and aero engine parts, automobile components and domestic equipment.

Factors affecting costs

Relative equipment cost (MMA=1) 10 – 30
Consumables – none
Mechanisation or automation
    Already mechanised
Skills required
    Operator – low
    Supervisor/maintenance engineer – moderate

Further information

American Welding Society: 'Welding handbook', Vol 3. Publ AWS, Miami, Fl.,
    AWS, 1980, 7th ed.

Resistance spot (RSP)

Principle

A resistance welding method in which the welding current is concentrated at the faying surfaces of overlapping sheets by cylindrical copper alloy electrodes pressed on the outer surfaces. A force is applied to the electrodes to squeeze the parts together and a current pulse is passed. Force is maintained on the electrodes and may even be increased following the current pulse to consolidate the weld and prevent cracking or porosity. Electrodes become heated and are cooled by an internal flow of water. The process is used on many different materials over a wide range of thicknesses from fractions of a millimetre up to about 5mm or even more. Power sources range from simple single phase AC with the electrodes in a single turn secondary to three phase frequency convertors and DC

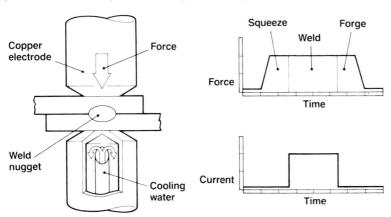

Principle of resistance spot welding.

machines. Access is normally required to both sides of the joint and where welds have to be made some distance from the edge of an assembly this requires a long reach or deep throat machine. Limits are imposed on the reach of a machine by the increasing loss of stiffness in the supporting arms and the fact that the presence within the throat of a large amount of magnetic material increases impedance and reduces current. A way round this difficulty is to use series welding in which both electrodes are placed on one side and current is conducted from one electrode through the sheets into a backing bar and back through the sheets to the second electrode. Two welds are then made at the same time. Access can often be improved by using cranked or off-set electrodes but these are mechanically weak and result in increased electrode wear.

The automotive industry formerly made extensive use of what is called 'multi-welding' in which a number of electrodes, each pair having its own transformer, are mounted in a press frame. This arrangement allows a complete component e.g. a door, to be welded at one time, although the electrodes may be energised in sequence. Such massive dedicated plant is becoming less common with the use of portable welders mounted on robots which permit reprogramming when a model is changed.

Method of use

The weld nugget is normally of slightly greater diameter than the electrode tip diameter which is usually (2.5 + 2s)mm, where s is the single sheet thickness. For the best quality welds the distance from the centre of the electrode to the nearest sheet edge should be not less than 1.25d, where d is the electrode tip diameter. Where sheets of unequal thickness are welded different size electrodes are used with the smaller diameter electrode placed against the thinner sheet. If dissimilar metals are to be welded a smaller diameter electrode is placed against the lower resistivity material. Where a continuous weld is required and seam welding is not possible a series of overlapping spots, stitch welding, (UK but not US term) can be used.

Advantages and limitations

The process can weld a variety of materials in sheet form and does not require consumables or edge preparation. Once set up little skill is required but machine maintenance must receive attention. Equipment is available from micro-size up to large multiple electrode installations. Weld times are usually only a few cycles duration and the process is highly productive. Joints must be lap or flange type and access has to be provided for the electrodes. The limited reach of machines may prevent welds being made in certain positions in large assemblies.

Applications

Miniature welders are used in the electronic and electrical industry. With larger press welders or portable plant the process is widely used in automobile body and component manufacture, for aircraft structures, engines, general engineering and domestic equipment.

Factors affecting costs

> Relative equipment cost (MMA=1) 1.5 – 20 or more depending on
>> tooling and whether or not automatic feeding is provided
> Consumables – none
> Mechanisation or automation
>> Already mechanised
> Skills required
>> Operator – none
>> Supervisor/maintenance engineer – moderate

Further information

American Welding Society: 'Welding handbook', Vol 3. Publ AWS, Miami, Fl., 1980, 7th ed.

Resistance upset (RU)

Principle

This process is closely related to flash welding using basically the same equipment. The parts to be welded are pressed together and the contact resistance creates heat on the passage of current. No melting occurs but as the contact area becomes plastic the joint is forged and this motion consolidates the joint by creating plastic flow at the interface.

Method of use

The type of joint usually employed is a butt in bar or section or between bar and section to a matching contour on a forged or machined part.

Advantages and limitations

The process is simple and clean to operate but with larger sized work is thermally less efficient than flash welding.

Applications

The process has been displaced for joints over about 150mm^2 by flash welding but is still used for welding wires, strip and small rods and in applications where cleanness and a smooth joint contour are required.

Factors affecting costs

Relative equipment cost (MMA=1) 1 – 5
Consumables – none
Mechanisation or automation
        Already mechanised
Skills required
        Operator – none
        Supervisor/maintenance engineer – moderate

Further information

American Welding Society: 'Welding handbook', Vol 3. Publ AWS, Miami, Fl.,
    AWS, 1980, 7th ed.

Soldering (S)

Principle

A low melting point filler metal (solder) is caused to flow under the
influence of heat between the parts to be joined. There is little difference in
principle between soldering and brazing but by definition the solder must
melt at less than 450°C. The joint must first be cleaned so the solder will
wet the surface of the parent metal for which purpose a flux is often used.
Superficial alloying between the solder and surface of the metal takes
place.

Method of use

There is a range of solder alloys with different melting points and
characteristics used with a variety of fluxes. Tin–lead alloys are the most
commonly used but the addition of antinomy (up to 6%) improves the
strength of these alloys. The tin–5% antinomy alloy has the best electrical
properties and good strength up to 150°C. Tin–silver alloys are also used.
The zinc–5% aluminium alloy is used on aluminium and some of its alloys
when soldering may be aided by the use of an ultrasonic iron or bath. Lap
joints, with preferably some mechanical connection such as a fold lock, are
preferred. Heat may be provided in a variety of ways e.g. by flame, heated
iron, induction, infra–red lamps or by dipping in a bath of molten solder
(and its variant 'wave' soldering).

Advantages and limitations

There is only a limited thermal effect on the workpiece and when the
appropriate heat source is chosen, soldering, which is capable of mechanisation, is a relatively fast method of joining. The joint strength is not
usually as high as for a brazed joint and the dissimilarity in colour and

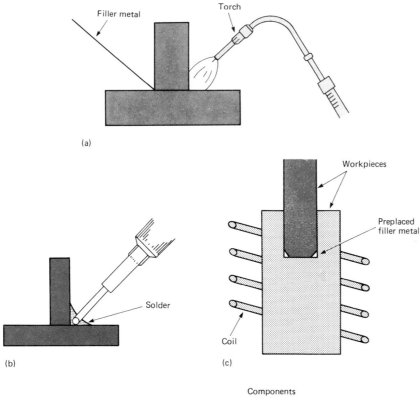

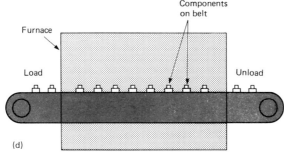

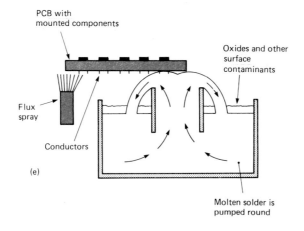

Five ways of applying solder: a) Manual, flame heating and hand fed filler metal; b) Manual, heating with soldering iron; c) Mechanised, induction heating; d) Mechanised, furnace heating giving continuous production; e) Wave soldering as used for printed circuit boards.

nature between bond and parent metal is unsuitable for some applications. Copper based alloys and mild and low alloy steels are readily soldered. Solders containing antinomy are not used on brasses. With steels the difficulty of soldering increases with carbon content and cast irons are not soldered. Stainless steels require special fluxes but can be joined with lead–tin solder.

Applications

Widely used in the electrical and electronic industries and for radiators, domestic pipe joints and fittings of all sorts.

Factors affecting costs

Relative equipment cost (MMA = 1)
    0.1 – 1 for simple systems, 10+ for highly mechanised plant
Consumables – solder and fluxes
Mechanisation or automation
    Can be mechanised
Skills required
    Operator – none/moderate depending on degree of mechanisation
    Supervisor/maintenance engineer – moderate, depends on degree of automation

Further information

Thwaites C J: 'Soft soldering handbook'. Publ International Tin Research Institute, 1977, 118 pages.
Manko H H: 'Solders and soldering'. Publ McGraw Hill, New York, 1979, 2nd ed. 350 pages.

Submerged-arc (SA)

Principle

An arc welding process in which the arc and weld zone are completely submerged under a blanket of granulated flux. A continuous wire electrode is fed from a coil through an electrical contact close to the arc. A hose feeds the granular flux from a hopper to a point close to the arc in the joint. Wires in the range 0.8 – 6.0mm diameter are usually employed and with wires at the lower end of this range (up to 2.0mm) a constant–potential DC power source can be used allowing arc length control by the self–adjusting effect. A power source of this type gives changes in welding current which alter the burn–off rate of the wire in a way which opposes changes in arc length thereby giving a steady arc length and voltage, (see also MIG welding). With the larger diameter wires a wire feed mechanism is required in which the rate of feed is controlled by the arc voltage to give a constant

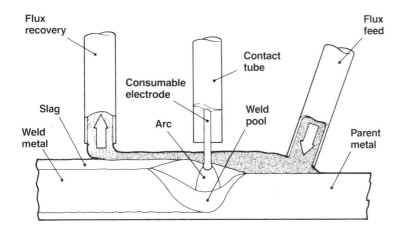

Principle of submerged-arc welding.

Submerged-arc head mounted on self-propelled, self-guiding tractor (courtesy ESAB).

arc length. This method of arc length control requires a constant–current power source which allows either AC or DC to be used.

With the smaller diameter wires it is possible to use the process manually but generally the process is mechanised with the welding head mounted on a tractor or positioned over moving work. Only welding in the HV or flat position is possible. Many variations of the basic single electrode process are possible the most common being the use of 2 or more wires or even strip electrodes. The process is used to weld carbon, carbon–manganese and alloy steels and stainless steels.

Method of use

The process is used on butt and fillet welds in the flat position and on fillets in the horizontal-vertical position. Single electrode submerged-arc machines may be mounted on self-propelled tractors carrying a flux hopper and the coiled electrode. A suction device may also be carried to recover the unused flux for reuse. Alternatively the welding machine may be moved over the work by hanging the machine from a gantry or boom. Cylindrical workpieces are generally rotated under a fixed welding machine. Multiple electrode or strip electrode machines are generally mounted on gantries or booms.

The high welding currents which are possible can give deeply penetrating weld beads allowing thick sections to be welded but allowance may have to be made for the higher dilution than with MMA, especially in root runs. Relatively narrow angled edge preparations can be employed and with the high deposition rates which are possible this makes the process highly productive.

The combination of flux and wire must be chosen with the application in mind e.g. the general purpose manganese-silicate flux for high welding speed and basic fluxes for clean weld metal and the best mechanical properties. When welding alloy and stainless steels the flux must prevent any alloy loss during welding. Special techniques give higher deposition rates by using metal powder additions or achieve higher joint completion rates by using narrow gap techniques.

Advantages and limitations

The process allows deposition rates and welding speeds greater than any other arc welding process and is highly productive. It can produce on the one hand deeply penetrating high quality welds and on the other may be used for weld surfacing where shallow penetration is desired. Existing in many forms, it can be adapted to allow the welding of material from a few millimetres thickness up to more than 100mm. The covering of flux suppresses fume and the light of the arc which allows the operator to dispense with a welding shield. The disadvantages which are not serious are that the weld pool cannot be seen, the granulated flux can get into machinery parts and all welding must be carried out in the flat or HV positions. Circumferential welds cannot be made in small diameters because the flux falls away.

Applications

Although used mainly on ferrous materials and stainless steel the process has been used on certain non-ferrous metals as well. It is widely used in shipbuilding, offshore, structural and pressure vessel industries, general fabrication, rebuilding and surfacing. Pipe welding using multiple wire techniques is particularly successful.

Factors affecting costs

Relative equipment cost (MMA=1) Single wire 3 to 5, more elaborate multi-wire, 10 or more
Consumables – electrode wire and fluxes
Mechanisation or automation
Semi-automatic but usually mechanised, can be automated
Skills required
Operator – none/moderate
Supervisor/maintenance engineer – moderate, depending on degree of automation

Further information

Houldcroft P T (ed): 'Submerged-arc welding'. Publ Abington Publishing, 1989,
    106 pages.

Thermit (T)

Principle

The process is a variation on the foundry technique of 'casting on' in which
superheated metal is poured into a preheated joint to fuse the parts
together. Heat is supplied by the strongly exothermic chemical reaction
between finely divided aluminium and the oxide of the metal required for
the joining process. This is usually iron or steel but can be a non-ferrous
metal such as copper. When steel is required the oxide is ferric oxide
(Fe_2O_3) mixed with small amounts of other metals to give the desired
chemical composition after the reduction process with aluminium.

$$Fe_2O_3 + 2Al = 2Fe + Al_2O_3$$

The charge, which may also contain scrap steel, is placed in a crucible
which can be bottom tapped by the removal of a pin and plug. The
chemical reaction is started by igniting magnesium powder on the surface
of the charge which is then rapidly converted by heat into aluminium oxide
and superheated steel. The steel so produced contains traces of aluminium.

Method of use

The parts to be welded are cut square and spaced apart in a mould where
they are preheated with a suitable burner. The crucible containing the
charge is arranged above the mould and when the reaction is complete and
the aluminium oxide has risen to the surface the underlying superheated

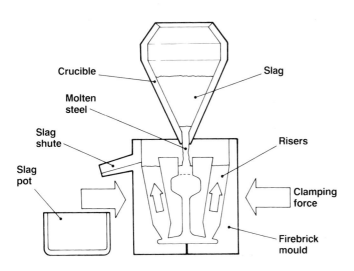

Method for welding a steel
rail, position at the end of
pouring steel and before slag
flows over into slag pot.

steel is tapped into the mould where it flows between the parts to be welded. When the metal has solidified the mould is broken open and the joint is fettled to remove excess metal.

Advantages and limitations

Electric power is not required so the process is suitable for site welding and for use in remote locations. It is rapid compared with other methods for welding heavy sections but can be difficult to control precisely. The mechanical properties, particularly toughness are not as good as with many other methods of welding.

Applications

Originally used for welding steel rails this is still the main use today. In the 1940s and 50s the process was used in shipbuilding and heavy construction as well as for repair welding and joining reinforcing bars. These uses are now rare.

Factors affecting costs

Relative equipment cost (MMA=1) 0.25
Consumables – Thermit powder, igniters and moulds
Mechanisation or automation
    Cannot be mechanised
Skills required
    Operator – moderate
    Supervisor/maintenance engineer – moderate

Further information

'Thermit welding'. Publ Thermit Welding (GB) Ltd, Rainham, Essex, 1963, 51 pages.

Tungsten inert gas or gas tungsten arc (TIG)

Principle

An arc process in which the arc is struck from a non-consumable electrode of tungsten to the work, the electrode, arc and molten pool being shielded by a stream of inert gas, usually argon. If filler metal is required it is added separately. The arc supplies heat for fusion of the work but also, when the electrode is positive, it exerts a cleaning action. DCEP, however, results in the electrode being overheated and when cleaning is required, as when welding aluminium alloys, AC is used. Cleaning then takes place on the

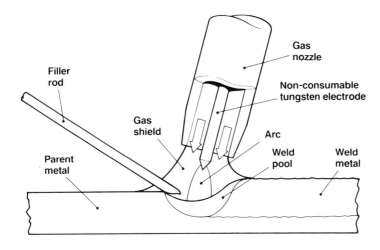

Principle of TIG welding.

Equipment for using TIG manually. Power source has facility for slope in and out of current (courtesy Murex).

Automatic tube welder set up on 100mm diameter pipe (courtesy NEI Weldcontrol).

electrode positive cycles. Modern power sources allow the balance of electrode positive and negative cycles to be varied so providing adjustable control of fusion and cleaning. Ferrous metals are welded with the electrode negative which gives greater penetration than AC. As it is undesirable to start the arc by touching the work because of the risk of contaminating the tungston electrode a high frequency discharge is used. With DC operation this may be switched out once the arc has started but it can be left on with AC to stabilise the arc as the current is reversed on each half cycle, passing through zero when the arc may be extinguished. Modern square wave power sources do not require arc maintenance of this type.

In pulsed TIG welding the current switches between a low background current which maintains the arc and a higher level at frequencies of several times/sec. Pulsed TIG welding gives improved control of penetration on thin metal and allows dissimilar thicknesses to be joined.

When mechanised welding thick metal or in surfacing, the deposition rate can be improved by using a 'hot wire' technique. The filler wire end is permanently immersed in the molten pool and current from a separate power source is passed through its last few centimetres raising its temperature by resistance heating and increasing the melting rate.

Method of use

TIG welding is a precision welding process which can be used manually or in a mechanised form. With suitable joints filler metal may not be needed and the arc can be used to melt down the joint, as in joint 270 for example. The short arc length required calls for manual skill or in mechanised welding an arc voltage sensor which controls the height automatically. Welds may be made both manually and automatically in most welding positions.

The fact that heating and filler metal addition can be independent makes the process particularly suitable for difficult tasks such as the root runs of unbacked welds and particularly for welds in pipe. Special portable automatic equipment is available for making pipe and tube butt welds automatically.

The weld metal in TIG welds is of high quality and there are few metals and alloys on which the process cannot be used. With reactive metals such as titanium the gas shielding is extended to cover the hot metal behind the weld pool.

Advantages and limitations

The major advantages are the quality of the welds and the precise control which can be obtained. Most metals can be welded. Mechanisation is easy and many specialised pieces of automatic welding plant are in use. Good penetration can be obtained but both penetration and deposition rate is much less than can be obtained with submerged-arc or MIG welding. It is not suitable for making fillet welds as good root fusion cannot be obtained. It is less suitable than other arc processes for welding thick metal.

Applications

The production of quality welds in a wide variety of joint types in almost any material but mainly for thin sections or precisely made joints. TIG is one of the most commonly used processes for dedicated automatic welding and is used in automobile, aerospace, power generation, process plant, electrical and domestic equipment manufacture.

Factors affecting costs

Relative equipment cost (MMA=1) 3 for simple manual equipment up to 10 for special plant.

Consumables – inert gas for shielding, replacement tungsten electrodes and filler metal

Mechanisation or automation
Manual but can be readily mechanised

Skills required
Operator – moderate/high
Supervisor/maintenance engineer – moderate but depends on degree of automation

Further information

Lucas W: 'TIG and plasma welding'. Publ Abington Publishing, Cambridge, 1990.
Cornu J: 'Advanced welding systems', Vol 3, 'TIG and related processes'. Publ IFS Publications, Bedford, 1988, 369 pages.

Ultrasonic (U)

Principle

Ultrasonic vibrations at about 25kHz produced by a magnetostrictive or piezoelectric transducer are applied to lap joints pressed between the transducer and an anvil. The ultrasonic vibrations produce localised slip and plastic deformation between the parts which rolls up oxide and contaminant films so creating a solid phase weld. The process can produce spot or seam welds or annular welds.

Method of use

Although high power ultrasonic equipment exists most applications are in foil and thin sheet. Both parts may be of the same thickness but dissimilar thicknesses are readily welded e.g. one may be thicker sheet or a shape such as wire. Aluminium is readily welded and because there is no fusion dissimilar metals may be welded.

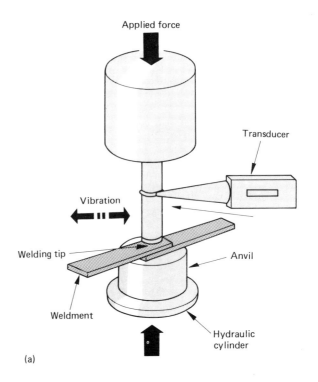

(a)

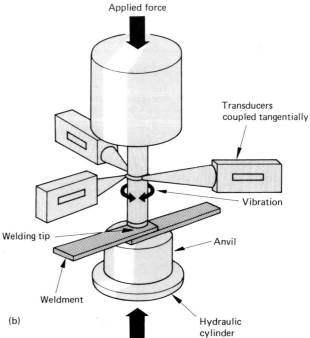

Ultrasonic systems for: a) Spot
welding; b) Ring welding.

(b)

Advantages and limitations

There is little deformation and only a local temperature rise so heat sensitive metals and components can be welded. Dissimilar thicknesses and materials are readily joined. Equipment costs escalate if thicknesses above 0.5mm are to be welded.

Applications

Electronic, small electrical and instrument manufacture.

Factors affecting costs

   Relative equipment cost (MMA = 1) 3 – 10
   Consumables – none
   Mechanisation or automation
      Already mechanised, can be provided with automatic feed
   Skills required
      Operator – none/moderate
      Supervisor/maintenance engineer – moderate but depends on degree of automation

Further information

Johnson K I: 'Introduction to microjoining'. Publ TWI, Abington, 1986, 34 pages.
Potente H: 'Ultrasonic welding – principles and theory'. *Materials and Design*, 1984 **5** (Oct/Nov) 228–34.

Learning Resources
Centre

HANDBOOKS IN WELDING TECHNOLOGY SERIES
Expert Guides for the Welding Professional

A GUIDE TO DESIGNING WELDS
John Hicks

Welding is rightly the responsibility of professional welding engineers. But the designer too needs an understanding of the principles and procedures involved if he is to brief the engineer or fabrication superintendent effectively.

Written with this design/welding interface in mind, John Hicks' book tells designers all they need to know about welding. Most importantly, it sets out what information should be given to the engineer or fabrication superintendent so that the designer's aims can be achieved, in terms of performance, safety, reliability, cost and appearance.

1 85573 003 0, 64 pages, photos and line drawings throughout, A5, paper

SUBMERGED-ARC WELDING
Edited by P T Houldcroft

This book provides an up-to-date guide to an important, versatile and highly productive welding process. Basically a simple process, in which an arc is struck from a continuously fed electrode wire to the work under a blanket of powdered flux, it was the first successful method of automatic arc welding.

The book is ideal as an introduction to submerged-arc welding. It will also inspire those with some knowledge of the process to look at the less widely used variants, many of which offer advantages in particular applications.

1 85573 002 2, 106 pages, photos and line drawings throughout, A5, paper

TIG AND PLASMA WELDING
Process techniques, recommended practices and applications
W Lucas

The TIG and plasma processes are used whenever high-quality welds are needed – for example in nuclear and chemical construction and in aeroengine components.

Here is an essential, practical handbook for designers, engineers and metallurgists, packed with all the information they need to make the best use of TIG and plasma welding in their own company.

As well as explaining the basics of the process and its applications, the author outlines the latest advances in equipment and operating techniques. He also describes the types of equipment – including transistor power sources and computer-controlled equipment – available commercially.

1 85573 005 7, 112 pages, photos and line drawings throughout, A5, cased.

AP Abington Publishing
Abington Hall, Abington, Cambridge CB1 6AH (Tel: 0223 891358)